Modern Astronomy: An Introduction

Modern Astronomy: An Introduction

HANS KIENLE

Translated by Alex Helm

FABER AND FABER LIMITED
London: 24 Russell Square

First published in 1963
as Einführung in die Astronomie
by R. Piper & Co. Verlag München
This translation first published in 1968
by Faber and Faber Limited
London: 24 Russell Square W.C.1
Printed in Great Britain
by William Clowes and Sons, Limited
London and Beccles

SBN 571 08725 6

Contents

Illustrations

Foreword

In the chapters which follow, my attempts to describe the growth of our concept of the universe, its structure and the laws which govern the phenomena which occur in it, claim neither to be a complete account of the history of astronomy in the twentieth century, nor even an objective report in the sense of an historian evaluating facts. Instead I wish to write about astronomical research during the first half of this century as I have experienced it myself, learning, investigating and teaching, and thus in some small measure also contributing.

The return of Halley's Comet predicted for the year 1910 marks the beginning of the period under review, while the launching of the first manned spacecraft closes it for the time being.

In that year in which Halley's Comet made its return two books were published in Germany; they were written in a style which was readily understandable, and summarised what was then known about the universe: *Aus Fernen Welten* (From Distant Worlds) by Bruno H. Bürgel, and a new edition of Littrow's *Wunder des Himmels* (Marvels of the Heavens) prepared by P. Guthnick. Many of the astronomers who shaped the destiny of the science during the second half of the nineteenth century were likely to have gained some early inspiration from Littrow's famous book which originally appeared in 1834–36. And Bürgel's book aroused in my generation of students an enthusiasm for the wonder-world of the stars.

1 · The Year of the Return of Halley's Comet

Let us consider for a moment what sort of ideas had grown up about the universe by the turn of the nineteenth–twentieth century. The nineteenth century may be considered as a period of inventory. It began with the discovery of the 'missing planet' between the orbits of Mars and Jupiter, which in fact turned out to be the first of the minor planets, one of many in this region of the Solar System. Then there followed the extension of the planetary system, the discovery of a new major planet, Neptune, on the basis of theoretical calculations. And once the distance to one of the stars had been established the Solar System could be integrated into a larger system of stars.

Comprehensive catalogues were compiled listing the exact positions of the stars, catalogues which were to form the basis for future calculations of the proper motions of these stars. Then there were also the catalogues of fundamental stars (position stars) for establishing the system of astronomical co-ordinates, while the lists of double stars, star clusters and nebulæ provided subject matter for advance beyond the limits of the Milky Way system. Catalogues of apparent magnitudes, colours and spectra of stars, lists of variables with periodic or irregular light changes, all these provided the foundations for the first analytical essays of the young science of astrophysics, whose prime subject was the Sun.

Fifty years after the discovery and interpretation of the Doppler Effect the first catalogue of radial velocities was published. With spectroscopic binaries, whose orbital motion can be deduced from periodically variable radial velocities, a new category of cosmic objects entered the scene.

Towards the end of the nineteenth century this extensive and painstaking inventory led to the development of new concepts of the structure of the universe, which were to influence and direct subsequent efforts. These concepts, in the form in which they eventually found their expression in the literature of the early years of the twentieth century, may be briefly outlined as follows:

The Earth, on which man lives out his life, is a celestial body which is a member of a system whose prime body is the Sun. The Sun itself is a star among stars, and the whole Solar System is but part of a much larger system of stars. This system of stars known as the Milky Way, or the Galaxy, belongs probably to a system of higher order, whose members are spiral nebulæ similar to that in the constellation Andromeda.

The ideas expressed by Kant in *'Allgemeine Naturgeschichte und Theorie des Himmels'* (General Natural History and Theory of the Heavens, 1755), and by Lambert in *'Kosmologische Briefe über die Einrichtung des Weltbaues'* (Cosmological Notes on the Structure of the Universe, 1761) that the universe consists of a hierarchy of several kinds of systems, were given a theoretical basis by the Swedish astronomer Carl V. L. Charlier of Lund in a paper entitled *'Wie eine unendliche Welt aufgebaut sein kann'* (How an infinite world may be built up, 1908).

In detail the picture of the universe as envisaged in 1910 was as follows: like the Earth, the planets Mercury and Venus within the Earth's orbit, and the planets Mars, Jupiter, Saturn, Uranus and Neptune beyond it, all travel round the Sun along paths which are very nearly circular, and which lie more or less in the same plane. The way in which the planets move is explained by the well-known Keplerian Laws, and these in turn are based on Newton's Law of the mutual attraction of masses (Law of Gravity).

The planets in their turn have satellites which orbit them, and

14

thus form sub-systems within the overall planetary system. The giant planets Jupiter and Saturn with their numerous moons seem to represent veritable replicas of the planetary system. Saturn is the possessor of a unique feature, the ring system which surrounds it; photometric and spectroscopic observations have shown the rings to be made up of very minute meteor-like particles.

As well as the major planets and their satellites, and the minor planets, which occupy the region between Mars and Jupiter, the Sun's family also includes comets, wanderers who make their appearances in our skies every now and then with no immediately apparent rhyme nor reason; so infrequent are the bright comets that they attract general attention, and even in our days, engender superstitious fear. Their long drawn out orbits extend over the whole Solar System in all directions, sometimes coming closer to the Sun than Mercury. On approaching perihelion, the point of the orbit nearest to the Sun, the head of the comet, which at first appears diffuse and nebulous, develops a tail which is always directed away from the Sun, is curved to a greater or lesser degree, and often changes shape rapidly, and thus demonstrates the influence of forces other than just gravity.

In contrast to comets whose progress can be followed for a period of weeks, if not months, shooting stars, or meteors, are no more than ephemeral phenomena, particles of 'cosmic' dust which occupies the space between the planets in such rarefied distribution that one can scarcely comprehend it. If these particles are very small they burn out when they enter the denser layers of the Earth's atmosphere ('shooting stars'). Larger lumps of matter, sometimes weighing several tons, leave a pronounced light trail before exploding; the fragments ('meteorites') fall to the Earth's surface where they can leave sizable craters.

At certain times each year we can observe meteor showers; they are predictable and bear out the assumption that the heads of comets are small clouds of dust, and that, as a result of frequent encounters with the gravitational fields of the major planets, this matter has gradually spread out along the orbit of the comet.

Yet another manifestation of interplanetary matter is Zodiacal Light. It seemed safe to say that the cause of this effect was the reflection of sunlight by dust particles, but there was division of opinion on whether this dust lay in a ring round the Earth (rather like the rings of Saturn), or whether the Earth were embedded in an extensive dust cloud surrounding the Sun.

As a whole, therefore, the Solar System presents many different facets, and at the same time remarkable orderliness in the way the bodies which comprise it are distributed and move. All the planetary orbits lie very nearly in the same plane, and any inclination to the plane of the ecliptic is so small that in this respect one might almost consider the system to be two dimensional. Only the comets and some of the minor planets form exceptions to the rule, but these bodies have so little mass even in aggregate as to be negligible.

All the planets and most of the satellites orbit their respective primaries in the same direction, which is also the direction in which the Sun and the planets rotate on their axes. Only at the outer limits of the system do we find any deviation from this rule, with the satellites of Saturn, Uranus and Neptune, and the comets are apparently subject to no particular law at all.

When one examines the distribution of mass, certain definite trends become apparent, which may not be clear at first sight. The numerical progression which has become known as Bode's Law expresses the distances of the planets from the Sun; it has occasionally been compared with Bohr's theory for the orbits of electrons in an atom. But Bode's Law is not something that follows simply from the laws of celestial mechanics, and, if it is not merely an extraordinary coincidence, the reason for it must lie in the very origin of the Solar System. The same is also true for the fact that the inner, Earth-like planets, Mercury, Venus, Earth and Mars, are all of relatively low mass, with mean densities in the order of four to five times that of water, whereas the outer planets, Jupiter, Saturn, Uranus and Neptune, are extremely massive, between 15 and 320 times the mass of the

16

Earth, and with mean densities roughly of the same order as water or ice.

The belt of minor planets, or asteroids, separates these two distinct groups of planets. It is still a matter of conjecture, whether, during the period of the formation of the planetary system, conditions in this region were such that no planet of one sort or the other could be formed, or whether a former planet in that zone subsequently disintegrated for some unknown reason.

The essential factor which emerges from this picture of the planetary system is that all motion is governed by the prime body, the Sun, in which practically the whole mass of the system is concentrated. Even the largest of the planets, Jupiter, has a mass which is equivalent to about only one-thousandth part the mass of the Sun, while all the other planets lumped together make up less than one ten-thousandth of the solar mass. For this reason it is relatively simple to compute the motions of bodies within the system by means of Kepler's Laws.

Newton's Law of general gravitational attraction explains how the planets mutually influence the pure Keplerian motion, which is due entirely to the Sun, to give rise to perturbations in the orbits. The chief problem which faced classical celestial mechanics was the question 'Can all the elements in the motions of the various planets and their satellites be completely and solely explained by Newton's Laws, or do there remain some differences between observation and theory which would indicate some shortcomings in Newton's Laws?'

Simon Newcomb compiled tables relating to the motions of the major planets, which he based on all the observational data available up to the year 1885; in a concluding report published in 1895 he summed up the unresolved factors in the movements of the four inner planets, Mercury, Venus, Earth and Mars. If one takes into account any errors which might arise in the actual observation, then there is only one area of dispute between observation and theory: the rotation of the major axis of the ellipse of Mercury's orbit. In theory the progression of this axis should be in the order of $1\frac{1}{2}°$ in 100 years, but in practice this value is too great by 43 seconds of arc, that is to say slightly less than 1%.

One could attribute this difference to unknown masses inside the orbit of Mercury, but the hypothetical intra-Mercurial planet (already named Vulcan) was never found. The most acceptable theory was that proposed by von Seeliger, who identified the perturbing mass with the dust cloud manifesting itself as Zodiacal Light.

It could also be argued that Newton's theory required some revision. In 1915 Einstein put forward a solution; on the basis of his theory of relativity he calculated the rotation of planetary orbits, which in the case of Mercury bore out the observed result.

This gave a new twist to the problem. Everything now depended upon obtaining exact data of the observed rotation, so as to test the accuracy of the Relativity Theory on the one hand, and the other laws of gravitation on the other. A critical analysis made in 1925 reached the following conclusion: 'So far there has been no indisputable solution to the problem of the motions of the four inner planets. The differences between the classical theory and observed data lie in the direction indicated by the Relativity Theory, but the evidence cannot yet be considered absolute proof of the accuracy of Einstein's gravitational theories.'

At that time, revision of Newcomb's calculations as a result of further and improved data was considered a necessary adjunct to the solution of the problem, but, in fact, this had to wait for almost thirty years, and became possible only with the introduction of electronic computers.

There was yet another problem; this time at the outer limit of the Solar System. Slight but perceptible perturbations in the movements of Saturn and Uranus, as well as observations of 'comet families', tended to point to the existence of one or more trans-Neptunian planets. Thus, similar to the procedure adopted by Leverrier in the case of Neptune, the orbits of two hypothetical trans-Neptunian planets were computed, and a search initiated. In 1930 the ninth planet of the Solar System was discovered and given the name Pluto; whether a tenth exists has not yet been settled.

Finally, irregularities in the movement of our Moon must also be mentioned; these eventually led to a reappraisal of methods for measuring astronomical time. From the observed apparent velocity of the Moon it could be shown that our measurement of time, based on the rotation of the Earth on its axis, is not entirely uniform.

On the question of the place of the Solar System in the universe as a whole, it was soon established that there was little fundamental evidence to go on. Only in the case of a hundred or so stars was their distance from us known with any degree of accuracy, by virtue of their apparent shift of position as a result of parallax. Some of the closer stars appear to alter their positions against the background of more distant bodies when they are observed from opposite ends of a given base line; in this case the longest available base is the major axis of the Earth's orbit, and this effect is known as parallax. Clearly, despite the importance of this method as a means of measuring the universe, it would be unrealistic to try to approach the task of ascertaining the distances of the hundreds of thousands of stars listed in the nineteenth-century catalogues by means of individual calculations, not to mention the millions upon millions of fainter stars which make up the Milky Way. A solution offered itself in statistical methods based on apparent brightnesses and proper motion.

The most likely key to the structure of the stellar system was the Milky Way, which stretches across the sky in an irregular sort of band, varying in brightness and width and more or less following the arc of a great circle. In a lecture to the 'Wissenschaftlicher Verein' in Berlin in December 1908, and entitled *Über das System der Fixsterne* (The System of Fixed Stars) Karl Schwarzschild gave the following impressive account:

'Astronomers have studied the general distribution of the stars relative to the Milky Way. The first thing one realises is that there is more to the Milky Way than just the comparatively narrow band of luminosity across the sky which can be seen with the naked eye. Certainly those areas of the heavens which are remote from the Milky Way are not rich star fields, and the nearer one gets to the Milky Way the more abundant stars become, even though they do not as yet become so numerous as to merge into a general shimmer. The immense clouds of luminosity which make up the Milky Way proper merely represent the culmination of this steady build up. So far as the distribution of stars is concerned, the significance of the Milky Way is not simply local, but universal.

'The fact that there is a connection between the Milky Way and the overall distribution of stars shows from the outset that the universe is an entity, that it belongs together. As von Seeliger puts it: Each and every individual star which we can see forms a part of the Milky Way system, whose shape is that of a flat lens. The horizontal diameter of this "lens", representing the direction of the band of the Milky Way, is about twice its thickness. The density of stars within this ellipsoidal shape is by no means uniform; a greater concentration of stars is to be found towards the horizontal plane, and towards the centre of the system, from which we ourselves are not too far removed. Therefore the reason why the sky appears to become increasingly crowded with stars the closer one gets to the Milky Way is because in this direction one is looking through a greater distance of star occupied space.

'The most wonderful thing about this idea is that in this way the Milky Way encloses the entire host of visible stars in one finite unit. It is also possible to estimate the approximate size of this unit: 20,000 light-years through the long axis of the system, and 10,000 light-years in thickness. The whole system lies like an isolated island in empty space, and, at distances which are very great in comparison with the dimensions of the system itself, there are other star systems forming other Milky Ways.

'The realisation that the whole system of visible stars is finite and secluded is a fundamental one, and marks the end of an epoch.'

Some idea of the immensity of the task which faces the astronomer, who wishes to understand the system not just as a three-dimensional, but also as an organic entity, may be gathered from the following comparison which Schwarzschild makes: 'Let us suppose that we are looking at the universe from some hypothetical vantage point so that 1 million kilometres is equivalent to 1 millimetre. Then on this scale the stars appear no bigger than pinheads on average something in the order of 100 km. apart. Nevertheless, despite these tremendous distances in comparison with the dimensions of the individual components of the universe, there is a definite coherence.'

The unity of the system of stars, which seems to be indicated by their spatial distribution, is also substantiated by observations of their proper motions, and they are quite definitely not 'fixed stars' as they were often called at one time. In actual fact, the stars are travelling through the universe at the same sort of cosmic velocities as the planets, but compared with the latter they are so far away from us that their movements become apparent only over much longer periods of time. At the first glance the proper motions of the stars seem haphazard and in all directions, but on closer examination certain definite laws emerge which Schwarzschild described as follows: 'One notices that in the main the stars tend to move away from the region of the constellation Hercules, and towards the opposite pole of the celestial sphere. . . . One reason for this is that the Sun, just as all the other stars, is also in motion through space, and that, in fact, the Sun is moving in the direction of the constellation Hercules. . . . The apparent motion of the stars which visual observation suggests, is borne out by spectrographic observations and the Doppler Effect. The stars in Hercules are approaching us at the rate of 20 km./sec., while in the opposite direction the stars are receding at the same rate; thus we may assume that it is in fact the Sun which is travelling at 20 km./sec.

'Consequently, if one wishes to determine how fast the stars are really moving, it is necessary to deduct any effect of the Sun's own motion, and the matter is one of some complexity; this is because the degree to which the Sun's movement influences that observed in a star depends on distance: the nearer

21

the star the more marked the effect. As the distance increases, this factor becomes more and more difficult to calculate, since for the most part the distances of the bodies in question are not known with sufficient accuracy. Conditions are so confusing that, until a few years ago, it was thought that the movement of the stars was quite irregular. . . .

'Only in recent years have positive results been achieved. If one looks at the Milky Way system and thinks in terms of an analogy to the planetary system, one might say that all the stars in the former rotate about the vertical axis in the same way as the planets about the Sun. However, this concept is not correct, since uniform rotation of this sort does not exist in the Milky Way system.'

It looks far more as if the stars move in planes parallel to the Milky Way, perhaps even about a central point of the stellar system, in orbits which are more or less circular, and no more than slightly inclined to the general plane of symmetry, but with much the same frequency in two opposed directions. The Milky Way appears to be 'an immense highway along which the stars prefer to travel, and on which they approach and move past each other, a road which is, moreover, parallel to one of the diameters of the Milky Way system'.

One theory, put forward by the Dutch astronomer Kapteyn, to explain this phenomenon assumed that the bodies in the stellar system form two distinct groups, whose movement relative to each other is rather like two swarms of dancing midges, and that we happen to be watching them at a time when they are just passing through each other. Direction and velocity are by no means uniform within each swarm; each swarm as a whole, however, is moving at a given speed and in a given direction relative to the other.

By reason of the extraordinarily loose cohesion of the swarms —pinheads 100 km. apart—neither actual collisions between stars of different swarms, nor yet appreciable disturbances of their proper motions are to be expected. Each star maintains allegiance to its own swarm, just as the midges in the example, and after their encounter both swarms will drift apart again without having undergone any real changes in their internal structures.

In order to preserve the idea of the organic unity of the stellar system, Schwarzschild put forward another suggestion to explain the systematic drift in stellar motion. He assumed the distribution of velocities of stars in the Milky Way to be not quite so haphazard as that of the molecules in a gas, and that paths parallel to the plane of the Milky Way predominate. Greater movement takes place along this plane than at right angles to it, rather like light in a double refracting crystal, where the velocity is greater in the direction of the principal axis, than perpendicular to it. Kapteyn's two swarms find no place in Schwarzschild's theory; from our point of observation, slightly to one side of the centre of the system, we see only the stars in our vicinity moving in the predominant direction parallel to the Milky Way plane.

Whichever interpretation one may put on the observations, it remains that the stars tend to follow one of two main courses, the 'apex' as the objective of solar motion, and the 'vertex' as the principal direction of stellar drift.

If all the stars make up one entity, a system of stars, that is to say the Milky Way system, then the question arises how one is to integrate other objects such as star clusters and nebulæ into this system. It was generally accepted that the loose clusters such as the Hyades, Pleiades and Præsepe, the diffuse gaseous nebulæ such as the nebula in Orion, and planetary nebulæ of the same type as the Ring Nebula in Lyra all belong to the Milky Way system; there was, however, some dispute about the nature of spiral nebulæ and elliptical nebulæ such as the Andromeda Spiral with its two companions, as well as numerous flattened nebulæ, which looked rather like spiral nebulæ seen edge on. Spectral observations showed that all these were not genuine nebulæ, but star systems. Yet so long as their distances from us remained unknown, little could be learned of their true dimensions.

All the arguments concerning the 'island universe' nature of the spirals were still fairly inconclusive around the year 1900.

23

In the paper by Karl Schwarzschild which was mentioned earlier he states, 'Even Herschel had already described the Andromeda Spiral as another Milky Way situated at an immense distance from us. Today we still consider this to be a likely explanation for spiral nebulæ: they constitute the only admissible exceptions to the axiom that all visible celestial objects belong to our Milky Way system.'

Bruno Bürgel suggested no more than tentatively that 'there may be Milky Way systems beyond our own Milky Way'.

The introduction to the recently published *Hubble Atlas of Nebulæ* begins with words to this effect: 'What are galaxies? No one knew before 1900. Very few people knew in 1920. All astronomers knew after 1924.'

2 · Elements of Progress—Instruments and Methods

The extent of our knowledge of the universe as a whole is very largely governed by the sort of instruments available. The discovery of the telescope marked the first great turning point, for, until this time, our knowledge was entirely confined to that which could be discerned with the naked eye. Improvements in optical equipment both in size and quality meant further advance. At the turn of the eighteenth–nineteenth century, William Herschel first put forward the suggestion that our Milky Way system was shaped rather like a double-convex lens. He based his ideas on the evidence of star counts made with the aid of reflecting telescopes whose mirrors he had ground himself; the largest of these was a mirror with a diameter of 48 inches, and a focal length of about $39\frac{1}{2}$ feet. With these same telescopes he discovered a wealth of star clusters and nebulæ, and compiled a comprehensive catalogue of the objects he had found. During the nineteenth century the nature of these phenomena became an engrossing and all-important problem. In 1838 F. W. Bessel succeeded in making the first stellar parallax measurement, which led to the first reliable determination of the distance of a star, by means of a heliometer; this instrument consists of a telescope whose objective has been cut down the middle, so that the relative positions of the two halves can be used to calculate angular distances.

An important stage in development was reached when the telescope was used in conjunction with other aids; objective measurement became possible, where there had formerly been but visual evidence. The addition of the spectroscope to the range of instruments which could be used with telescopes began a new era of astronomy based on the analysis of starlight.

At about the same time photography began to make an impact on astronomy. In 1841 John W. Draper at the Harvard Observatory managed to take the first successful photograph of the Moon using a Daguerre plate. In 1857 at the same observatory William Bond produced the first astrophotographs by means of wet collodion plates. In England William Huggins photographed the spectrum of Sirius, while in Germany Hermann Vogel replaced the spectroscope, which until that time had been used to determine radial velocities by means of visual measurements of line shift in the spectrum, with the spectrograph, thus making the method about ten times more accurate.

Photographic plates have a number of advantages over the human eye. In the first place, the plate has the ability to accumulate and to preserve impressions of light, and so, given a sufficiently long exposure, a photographic plate is capable of recording objects too faint to stimulate the eye. Another advantage is that a single photograph will record simultaneously all objects in the field, each in its correct position, whereas an observer at the eyepiece has to fix the position of each star individually in order to make his calculations. A photograph provides us with a document of conditions at a particular moment of time, whose testimony is complete and permanent, and to which reference may be made at any time.

Photographic emulsions can react to certain wavelengths of the electromagnetic spectrum to which the eye is insensitive. The 'actinic rays', as photosensitive radiations were at first called, range from blue, through violet and ultraviolet, to fast X-rays. With special sensitising media incorporated in the emulsion, sensitivity can be extended into the long-wave range of the spectrum: ortho- and panchromatic plates are sensitive to visible light, while infrared plates will capture longer-wavelength radiations which cannot normally be seen.

The usual silver bromide emulsion which is sensitive to the blue section of the spectrum will 'see' stars which emit mainly this kind of radiation as brighter than the human eye will see them; this is because the eye is more sensitive to light of longer wavelengths, towards the yellow end of the spectrum. The difference between the magnitude of a star as indicated on a photographic plate and its visual brightness yields a factor known as the colour index, which is not only a guide to the colour of the star, but also to its surface temperature.

So manifold are the possibilities of photography that it has found successful application in almost all branches of astronomy. In some respects the enthusiasm and optimism during the last twenty-five years of the nineteenth century were so great that good intentions sometimes out-stripped practicability. That much vaunted collective venture, the 'Carte du Ciel', a photographic map of the heavens, in which numerous observatories in all parts of the globe were concerned, never managed to achieve all that its initiators had hoped. Human failure was not taken sufficiently into account, and the capabilities and efficiency of new methods, as well as instruments, had to some extent been overestimated.

But the next generation learned from these mistakes; it recognized that there were limits to the photographic method, and also what these limits were. The relationship between exposure time and density was studied and eventually provided a formula for the density of photographic plates. Improvements were made in photographic materials, and their properties adapted to meet the great variety of practical requirements.

Basically there are two types of optical telescope available to the astronomer, the refractor and the reflector. In a refractor the optical image is produced by the refraction of light rays in their passage through glass lenses; the object glass of a refractor is a lens or a combination of lenses. In a reflector the image is formed by the reflection of light rays from the surface of a concave mirror. The real image produced by the object glass (or mirror, as the case may be) can then be examined directly through a magnifying eyepiece, or photographed, or directed

into some special device such as a photometer or spectro-
meter.

A simple lens, which is produced by grinding a piece of glass
into the required shape, gives an image surrounded by a coloured
halo, particularly if the object is bright. This happens because
different wavelengths of light undergo varying degrees of refraction in their passage through the glass. Blue light, which has
a shorter wavelength, is more sharply deflected than red light,
whose wavelength is longer; consequently a simple lens brings
red light rays and blue light rays to different focal points, and
similarly with other colours. Such colour aberration can to a
large extent be obviated, though never entirely eradicated, by
the use of compound lenses made from various types of glass
with differing dispersion properties; lenses of this kind are said
to be achromatic.

A Newtonian reflecting telescope, unlike a Keplerian refractor, does not suffer from colour aberration. On the other hand
this advantage is offset by the fact that only part of the resultant
image is geometrically true. A parabolic mirror brings only
those light rays to a sharp focus which lie parallel to the optical
axis of the mirror. As a result there tends to be some loss of
definition towards the edges, and only in a small field at the
centre are the stars true points of light.

There are also certain technical aspects which influence the
choice between refractor and reflector. Large mirrors can be
made more easily than large lenses; in the former the quality of
the actual reflecting surface is the important factor, while in the
latter the entire lens has to be homogeneous, that is to say that
the glass from which it is made has to be of uniform density and
free from flaws, since the light has to pass right through it. The
refractor of the Lick Observatory (91 cm. and built in 1888), and
that at Yerkes Observatory (102 cm. and built in 1897) may be
said to be the maximum diameters to which one may safely
aspire; 60–70 cm. is nowadays considered to be about the optimum size.

In the case of a reflecting telescope the limiting factor tends
to be financial rather than the ability to produce a large enough

28

block of glass from which the mirror is to be ground. Technological progress in the course of this present century has brought with it a steady raising of the limit. The 90-cm. Crossley Reflector of the Lick Observatory in California, and the 72-cm. Waltz Reflector at Heidelberg led the way into this century; the largest telescope of this kind to date is the 200-inch Hale telescope on Palomar Mountain. At present a diameter of about 200 inches probably represents the optimum size for reflecting telescopes; if one exceeds this figure by anything more than a small margin, the resultant increase in light-gathering power would be disproportionally small in the face of the extra expense involved. Indeed, one might almost say that such a project would be undertaken more to advance national prestige rather than science.

A special form of refractor had to be developed for astrophotography. The image field, even of an astrograph fitted with a double-lensed objective, is domed; consequently in the process of projecting the stellar images on to a flat plane, there is too rapid a loss of definition towards the edges. It is possible to overcome this difficulty by employing combinations of three or more lenses made of glasses of differing powers of refraction. And so for the purpose of photographing extensive star fields, astrographs fitted with triplet objectives by Petzval or Tessar took the place of ordinary astrographic refractors.

The Bruce Telescope at the observatory on the Königstuhl, fitted with a pair of such object glasses of 40 cm. aperture and 2 m. focal length, was used with great success by Max Wolf in the quest for minor planets, and also for studying dark nebulæ in the Milky Way and some clouds of spiral nebulæ. The number of nebulæ detected on plates taken with the Bruce Telescope, and listed and described in the Königstuhl Nebula Catalogue far exceeds that which William Herschel was able to find with his 20-foot long telescope. Astrographic plates which had been given long exposures provided data for star counts on which the investigation into the structure of the universe rests.

By putting a prism in front of the objective of an astrograph its field of application can be increased. Such a device is called

29

an objective prism camera, and is used for obtaining stellar spectra. The Henry Draper Catalogue, which contains details of the spectra of 225,300 stars, and which was compiled around the turn of the century at the Harvard Observatory, is based on the work of Miss Cannon on spectra obtained with just such an instrument.

The rival claims of reflectors and refractors are more or less equal, since each type plays its own special rôle in astronomy. The usefulness of a particular type of telescope for a particular task may be assessed on the following considerations:

1. The focal length determines the size of the image, and, together with the diameter, the resolving power, that is to say the ability to separate objects in close proximity.

2. The effective aperture determines the light-gathering power of the instrument, and thus the distance it can penetrate.

3. It is possible to construct multi-lens objectives in order to reduce chromatic errors and to correct the field; parabolic mirrors are free from colour aberration, but give complete light concentration only in the direction of the optical axis.

Stellar measurements requiring particular precision, as for instance in the case of double stars and the determining of parallax, are the domain of long-focus refractors or astrographs with wide apertures. Spectrographs with great dispersion, used for measuring radial velocities or line intensity, are used to best advantage in connection with reflecting telescopes. The refracting telescope is the principal instrument for astrometry, while the reflector rules the field of astrophysics. Astrographs stand about half way between the two; they can be used for precision measurements as well as for photographing stellar spectra when fitted with objective prisms.

The ideal mount for a telescope is the parallactic or equatorial type; this allows the telescope to follow the diurnal motion of the heavens by rotating it about one axis which is set parallel to the Earth's axis of rotation. The parallactic mounting of instruments with long focal lengths (in excess of 10 m. or so) presents some difficulty. In order to achieve the necessary stability and accuracy of guidance for very long telescopes, special types of mounts have been developed, in which the actual telescope is held rigidly in a vertical or horizontal position, or even inclined at an angle so that it lies parallel to the Earth's axis. Here the guidance is effected by means of a 'cœlostat', which consists of a series of clock-driven mirrors which direct the light from a particular part of the sky into the telescope proper.

A horizontally mounted telescope is normally used for observing the Sun during a total eclipse. At such times temporary observing stations are set up in suitable locations, and this method of mounting the instrument presents the least amount of technical difficulty. In the case of a more permanent solar observatory a tower telescope is preferable, because the vertical arrangement of the light path is then less susceptible to thermal agitation of air masses in the vicinity of the instrument; this factor is a distinct disadvantage in the horizontal mount.

Briefly, the equipment of a typical observatory at about the turn of the century consisted of a telescope with position circles; clocks; a transit instrument, meridian or vertical circle; a long-focus refractor, either for visual or photographic observation, or, as a combination of both; an astrograph (usually in the form of a twin astrograph); and a reflector designed for viewing at the Newtonian, or Coudé, or Cassegrain focus. In 1913 the newly established observatories at Hamburg-Bergedorf and at Berlin-Babelsberg are typical examples. They show us how the observers of the day proposed to tackle the problems which faced them; what instruments they considered most suitable so far as the technology of the day was able to provide them. In addition there might well be a tower telescope with all the accessories for observing the Sun.

If we now turn our attention to the question of how instrumentation has developed and improved since that time, and what findings influenced the direction of this development, one cannot help but think first of the several giant telescopes to be set up in California. In 1906 the 60-inch reflector on Mount

31

Wilson came into operation, the instrument with which Harlow Shapley broadened the horizons of the Milky Way system. In 1917 there followed the 100-inch Hooker telescope, with which Edwin Hubble studied distant galaxies, and discovered the significance of the red shift in their spectra, a fact which subsequently led to the concept of an expanding universe. The latest of the giants at the moment is the 200-inch Hale telescope on Palomar Mountain (1948) which has fulfilled and even exceeded all that was expected of it; with this instrument, man has been able to probe deeper into space than ever before.

Other observatories eventually followed the example of the Mount Wilson Observatory: in 1919 the Astrophysical Observatory in Victoria (Canada) obtained a 74-inch reflector; in 1939 the Yerkes Observatory set up a sister station, the McDonald Observatory in Texas, equipped with an 82-inch reflector. Both these instruments are used primarily for spectrographic work and have yielded astrophysicists a wealth of valuable information.

In Europe projects of such magnitude could be put into effect only since the end of the Second World War. In 1958 the 76-inch reflector in Haute Provence, France, and in 1960 the 79-inch reflector at the Karl Schwarzschild Observatory at Tautenburg near Jena, Germany, came into commission. Also during this time a 118-inch reflector was set up at the Lick Observatory in California (1959), and a 102-inch reflector in the Crimea, U.S.S.R. (1961).* Meanwhile a 98-inch reflecting telescope has been installed at Herstmonceux in Sussex.

In the field of astrographics, four-lens objectives took the place of triplets; these reproduce fields up to 30–40° so well, that with their aid a number of astrometrical problems can be solved with considerable accuracy. They have made it possible to obtain observational data required for determining proper motion; this is done by repeating photographs at intervals of no more than a few decades, so that it is no longer necessary to refer to visual observations made in the nineteenth century. The basis

*A 236-inch reflector is currently being constructed in the U.S.S.R.

for research into the dynamics of the universe has thus not only been enlarged, but also strengthened.

The happy inspiration of that lone-wolf in practical optics, Bernhard Schmidt, who was permitted to establish a modest workshop at the Hamburg Observatory, led to the conception and birth of an entirely new kind of reflecting telescope. He managed to solve the problem of how to increase the effective field of the curved mirror in a remarkably simple manner. He combined a spherical mirror with a specially ground, thin glass plate placed at the centre of curvature. After Richard Schorr, the Director of the Hamburg Observatory, had displayed photographs taken with the first Schmidt Camera at the congress of the Astronomical Association in Göttingen in 1933, this type of instrument received world-wide acclaim.

With a Schmidt prism camera of 61 cm. aperture the Warner and Swasey Observatory in Cleveland, Ohio, has been engaged in an ambitious programme of spectral study since 1941, especially in the red to infrared range of the spectrum. The 'Big Schmidt', which has been in operation at Palomar since 1949, is twice as big; this is the instrument which furnished the photographs for the *Palomar Sky Survey* showing all stars down to magnitude 20–21 on plates 18 inches square (equivalent to $8°.6 \times 8°.6$) taken through red and blue filters. The fact alone that this atlas, which runs to almost 600 pages, contains roughly ten thousand times as many stars as the plates in the *Photographic Star Chart*, which includes objects in the order of magnitude 14, not to mention the unimaginable wealth of other cosmic objects including far distant extra-galactic star systems, shows to what extent improved instrumentation has advanced the science of astronomy.

At about the same time as Schmidt was developing his revolutionary telescope, Maksutov in the U.S.S.R. brought out another type of 'coma-free' reflector; he combined a spherical mirror with sharply curved meniscus lenses. With such a 'Maksutov Telescope' Shajn and Hase at the Simeis Observatory in the Crimea obtained photographs of gaseous nebulæ in the Milky Way showing their filament-like appearance in such detail, that it far surpassed anything which had hitherto been obtained with a wide-angle lens.

The French astrophysicist Bernard Lyot had much the same impact in his particular field, which was the study of the Sun, as Schmidt in the field of astrophotography. With the 'Coronagraph' which he had developed it became possible to observe the solar corona without having to await the occasion of a total eclipse. In high altitude observatories, where the sky is a deep blue, the coronagraph and the spectroheliograph are important tools for the study of the Sun.

In addition to such purely optical instruments as refractors, astrographs and reflectors of various designs, together with their numerous accessories, such as photometers and spectrographs, for analysing the light from the stars, all of which are of necessity restricted to a range of wavelengths between 3,000 and 8,000 Å, an entirely new concept of telescope has entered the field since 1945. This is the radio telescope, by means of which the observable range of radiations has been considerably extended to include wavelengths measured in centimetres, decimetres and even metres.

Radio astronomy owes its origin to a chance discovery during research into atmospheric interference in wireless communication; the discovery was made by K. G. Jansky in the thirties, and then further developed during the war years in connection with radar. For obvious reasons much of what was being done at that time had to remain secret, and when at last the facts could be released the new data poured upon the world of science like an avalanche; it was quite unprecedented. The existence of cosmic radio sources was a firmly established fact, and much of the wartime high-frequency apparatus was made available to astronomers. Consequently a great number of very active research centres sprang up almost overnight, and new discoveries followed rapidly one on another. Eventually, adapted wartime radar equipment gave way to radio telescopes designed purely for astronomical work, more powerful instruments of larger dimensions, or specialising in certain wavelengths or on certain objects.

34

At the beginning of this epoch it was the physicist who was brought in to collaborate with the astronomer, above all the spectroscopist, who knew how to read and interpret the spectra of the stars. Now engineers and radio experts have made their entry into the field; in fact, they carried out the initial developments of radio astronomy almost entirely on their own. The stage has been reached when more and more cooperation between radio experts and astronomers is needed, the one to look after the technical side, and the other, using his knowledge of the universe, to direct the programme of research and to evaluate the data obtained. An example of the success of such teamwork is the discovery of the 21-cm. wave emissions of interstellar hydrogen gas. These had been predicted by the Dutch astronomer van de Hulst in 1945. Then in 1951 different observers in the U.S.A., the Netherlands and Australia furnished nearly simultaneously evidence to substantiate van de Hulst's theory.

The Earth's atmosphere is the greatest obstacle in the field of short-wave radiation investigation; the limit is somewhere around 3,000 Å. The Earth's atmosphere is also responsible for loss of clarity and unsteadiness of image in optical telescopes, since air masses of varying densities are constantly on the move. This is particularly noticeable in solar work, which naturally enough takes place during the day. As a result of these turbulences it is impossible in practice to obtain the degree of resolution in optical telescopes which might be possible in theory. To the astronomer, therefore, the atmosphere is a nuisance, and every possible effort is made to minimise its effect. The first step towards this end was the location of observatories in high mountains above the denser layers of the atmosphere. There followed attempts to reach higher altitudes by means of balloons. In the thirties Regener used unmanned recording balloons to lift spectrographs to a height of more than 30 km., and in this way managed to advance our knowledge of the ultraviolet end of the solar spectrum a little further. However, these balloons were not able to penetrate the ozone layer which is situated between 30 and 40 km. above the surface of the Earth, and is largely responsible for curtailing the spectrum. Then in 1945 U.S. technicians using V2 rockets captured from the Germans sent recording instruments higher than the ozone layer, and in 1946 eventually obtained records of the ultraviolet spectrum of the Sun to 2,100 Å. Meanwhile, observation has been extended right into the X-ray range.

Yet another valuable contribution is that of Martin Schwarzschild of Princeton. He uses a balloon-borne telescope called a stratoscope (his latest carries a reflector of 36 inches diameter) which is controlled from a ground station. With this instrument Martin Schwarzschild has obtained photographs of the surface of the Sun, which are free of any distortions through unsteadiness of the Earth's atmosphere. The sequence of photographs taken during a single ascent in 1959 provided a solution to the true size of the apparent granulation on the surface of the Sun, that is to say the size of turbulences in the solar atmosphere. For some years this had been the subject of much conjecture.

Artificial satellites represent the latest stage in instrumental development; they can carry recording apparatus well beyond the limits of our atmosphere, and then relay their information to a convenient ground station. Completely automatic space laboratories with television-camera-telescopes are projects of the not too distant future, and will provide astronomers with permanent observatories free from atmospheric interference.

Let us now consider briefly methods of observation and the evaluation of data. The trend may best be summarised in the terms 'objectivity' and 'automation'.

The introduction of photography represents the first step towards greater objectivity, if we take this to mean the exclusion of purely sensory perception. The reaction of light on a photographic plate provides an objective document, but it is still necessary to understand the language in which this document is written, if it is to yield any information. The scope and reliability of the information gained from a photograph are themselves determined by the scope and reliability of our knowledge of the laws which govern the way in which light affects photographic emulsions.

Photographic photometry, that is to say the measurement of stellar magnitudes by means of photography, is only objective in a restricted sense, since the blackening or density of the plate is not solely dependent upon the total amount of energy which reaches the plate during exposure. A short exposure at high intensity produces a different result from long exposure at low intensity, and it also makes a difference if the overall exposure time is, in fact, the sum of a number of short exposures at certain intervals, rather than continuous. This means that before a photograph can be used for photometric purposes the relevant 'characteristic curve' has to be determined.

As a result of the introduction of the photo-electric cell into astronomy by Paul Guthnick and Hans Rosenberg in Germany and Kuntz and Stebbins in the U.S.A., there came a new development in 1913, which from the outset raised the accuracy of measurement to ten times that which is possible by the direct observational method; it led to the gradual replacement of the photographic plate by the electronic multiplier, and the image converter, as well as other kinds of electronic aids such as those in use today.

The romantic idea of astronomers as star-gazing, nocturnal creatures watching the panoply of the heavens through the eyepieces of their great telescopes, searching perhaps for a comet, perhaps for some new star or planet, is now no longer true; it is like confusing the alchemist of old with the chemist of today. The desired, and in some cases already achieved, goal in instrumental development is a telescope which will furnish the astronomer not with a visual image but with a punched card or computer programme tape.

3 . Space and Time: the System of Astronomical Coordinates

Newton's concept of celestial mechanics assumes that both space and time are absolute; although this assumption was adequate as a starting point, it is not true. The definitions that Newton gives are as follows:

1. By virtue of its nature and without reference to any external factor, absolute space remains constant and static.
2. Absolute time, true, mathematical time, passes in the abstract, and, by virtue of its nature, uniformly, and without relationship to anything outside itself.

Celestial mechanics describes the movement of a body by defining the positions of that body at successive intervals of time.

Position in space is given by dimensions in three planes at right angles to each other, and, where they intersect, they realise the axes of a right-angled, three-dimensional system of coordinates. The absolute system of coordinates of classical celestial mechanics is as abstract an idea as Newton's absolute space on which its foundations rest. The system of coordinates on which observation relies in practice is concerned with definite objects, and for the astronomer the fixed points in space must be the stars; the empirical system of coordinates thus derives from a system of stars. Similarly, theoretical absolute time is also an abstraction. Empirical time relies on certain periodic recurrences; the astronomer's clock is the Earth rotating on its axis.

The transition from the empirical system of coordinates of practice to the fictional absolute system of theory demands a knowledge of their mutual relationship. This relationship is realised by means of a theory concerning the motion of the body with which the empirical system of coordinates is closely connected. This movement is composed of displacement of the zero point and the rotation of the directions of the coordinate axes.

Primarily, an astronomer can determine only the apparent position of a star in the sky, that is to say the direction from which the light from that star seems to come. Direction finding, however, calls for angular measurement. The two angles which determine the location of a point on the celestial sphere are said to be its spherical coordinates. The third factor, which finally fixes the object at one particular point in space, is how far the object is from the observer. This can be found by means of triangulation, by observing the object from either end of a given base line and noting the angles. In astronomy this kind of range finding through triangulation is known as 'measuring the parallax'. The word parallax means 'shift' and is the difference in direction of a distant object as seen from different places.

Thus the 'diurnal parallax' of the Moon is the shift in position which the Moon seems to undergo when viewed simultaneously from two places on the Earth which are one Earth radius (6,400 km.) apart. The 'annual parallax' of a star is the apparent shift of the position of the star as a result of the Earth's journey round the Sun; more accurately this is the angular value of the semi-major axis of the Earth's elliptical orbit as seen from that star. Then there is 'secular parallax', which denotes the shift in a star's position resulting from the fact that the entire Solar System is moving through space.

The primary system of coordinates is closely bound up with the Earth, so that it participates in the daily rotation of the Earth on its axis, as well as the latter's annual revolution round the Sun. As a result of the Earth's rotation about an axis which maintains a constant inclination of $66\frac{1}{2}°$ to the plane of the Earth's orbit throughout its journey round the Sun, one direction in space is clearly defined: this is the celestial pole around which all the stars appear to travel in concentric circles.

The orbital motion of the Earth round the Sun fixes a basic plane in space, describing a Great Circle along which the Sun appears to make its journey over the year; this apparent path of the Sun is called the ecliptic. The intersection of the plane of the Earth's equator and the ecliptic gives us the second axis of the empirical system of coordinates; the point is defined by the vernal equinox, or the First Point of Aries.

The celestial pole and the vernal point are the reference points for spherical coordinates. There is no particular mark in the sky which may be identified with either of these points, and their positions relative to the stars which form the framework of the empirical system has to be deduced from observation of the apparent diurnal rotation of the heavens on the one hand, and from the annual apparent movement of the Sun on the other. Such observations show that both the celestial pole and the vernal point are gradually changing relative to the background of stars.

From the point of view of celestial mechanics, the Earth resembles a top, which is spinning on its way round the Sun with its axis of rotation tilted at a given angle to its path. A top, however, begins to wobble when external forces act on it; its axis of rotation alters its direction, and 'precession' sets in. In the case of the Earth, the disturbing influences are the Sun and the Moon. Consequently the apparent celestial pole circles the pole of the ecliptic once in 26,000 years. At the same time the Moon causes the axis of the Earth to oscillate slightly (nutation) with a period of nineteen years.

Even the ecliptic does not maintain a constant position in space, but alters slowly because of the gravitational pulls of the other planets. This causes additional movement of the equinox.

Lastly, the fact that the velocity of light is finite means that it is subject to apparent aberration, so that the direction of the light coming to us from a given star is not absolutely the true direction to that star; the extent of this aberration depends on the velocity of the observer relative to the speed of light.

All in all, the stellar coordinates determined by the astronomers, as compared with those of the absolute system of celestial mechanics, tend to be distorted by a series of effects. By combining observation with theory, spherical astronomy with celestial mechanics, the sum of these effects, 'the reduction of the apparent position on the sphere to the true position', can be worked out in a fundamental system. Nevertheless, this can only be done by means of a process of approximations, since the accuracy with which the reduction can be calculated depends on the accuracy with which the necessary parameters are known at the time. Each improvement in the values of these constants, which it has been necessary to make in the course of time as a result of new observations, entails corrections of the coordinates and movements based on the fundamental system.

This fundamental system of coordinates is defined by the location of the equator and the ecliptic at a particular point of time in relation to a number of chosen 'fixed' stars, which make up the 'fundamental catalogue'. Since the stars of this fundamental catalogue are not really fixed in the strictest sense of the term, but are themselves moving individually along certain courses, their location in space is changing. Hence the system of coordinates which they support also changes from the original.

On the basis of the observations of the eighteenth century, Friedrich Wilhelm Bessel of Königsberg compiled the *Fundamenta Astronomiæ*, the fundamental catalogue which served the astronomers of the nineteenth century as their system of coordinates. Around the turn of this century the *Fundamentalkatalog des Berliner Jahrbuchs* came into being as a result of the intensive survey of the heavens undertaken by the astronomers of the nineteenth century. The man behind this catalogue was Arthur von Auwers, who set up a mathematical centre at the Berlin Observatory; this subsequently became independent as the 'Astronomisches Recheninstitut'. Also he began collecting material for a card-index at the Academy of Sciences listing the positions of stars ('Geschichte des Fixsternhimmels').

One of the traditional tasks of the Recheninstitut, which after the collapse of Germany in 1945 moved its headquarters to

Heidelberg, is to continue to improve the fundamental catalogue. Auwers himself compiled the *Neuer Fundamentalkatalog*
(NFK), which replaced the earlier work (FK) in 1907. August
Kopff, appointed to the Directorship of the Berlin Rechin-
institut from Heidelberg in 1924, brought out a third revision
(FK3) in 1937–38, and in the difficult war and postwar years
prepared the ground for FK4, which his successor Walter Fricke
is now publishing under the auspices of the Heidelberg Academy
of Sciences.

In accordance with a resolution passed by the International
Astronomical Union (IAU) in 1955 the *Fundamentalkatalog des
Berliner Jahrbuchs*, in which the number of stars listed has risen
from 925 under Auwers to 1,587 in the latest edition, now forms
the basis for all astronomical position determining; it represents
the obligatory international system of astronomical coordinates.

To what extent is this system absolute in terms of classical
celestial mechanics? In order to answer this question, we should
once more examine how each of the two systems, which we are
comparing, is to be defined.

The empirical system is determined by the positions of the
equator and the ecliptic at a particular time, let us say 1950·0
for FK4, relative to the positions and individual motions of the
fundamental stars, and the constants required for reducing
apparent to true position; the relevant constants are precession,
nutation, aberration and parallax.

The theoretical, i.e. absolute, system of coordinates is only an
abstraction based on the claim that within its framework the
laws of mechanics are valid. Classical mechanics are founded on
the law of inertia—Newton's first law: a body will remain at
rest, or move uniformly in a straight line, so long as no other
forces act on it. Consequently the theoretical system of coordinates is also known as the inertial system.

The laws of mechanics which an astronomer is best able to
test are primarily those which govern movement within the
planetary system, that is to say the motions of the planets and
their satellites. This brings us back to the question raised in the
first chapter: Can all the finer details of the motions of the

planets and their satellites be wholly explained by Newton's law of gravity, or, when one compares observation with theory, do certain considerations arise which indicate a divergence from the exact truth of Newton's equation?

To put this more finely: if differences between observation and theory do occur, then it has first to be shown to what extent such differences can be attributed to a shifting of the empirical system, which is based on fundamental stars, in contrast to the inertial system. The possible rotation of the empirical system of coordinates was actually considered in attempts to explain the movement of Mercury's perihelion.

On the other hand, apparent deviations from theory could have quite another origin. If the inertial system is introduced as a system of coordinates in which the mechanical laws of motion are valid, then it must be remembered that time features in these laws as an independent dimension. The laws of planetary motion, which were evolved from the law of gravity, give positions relative to absolute time in the Newtonian sense, i.e. based on an inertial system. Observations based on the fundamental system give such positions in relation to empirical, i.e. astronomical, time, which is based on the rotation of the Earth; the mean second of time, which the physicist also uses, is a $24 \times 60 \times 60 = 86,400$th part of the mean solar day.

For the purpose of measuring astronomical time, however, any other periodic celestial motion can equally well be used; for instance the revolution of the Moon round the Earth, or the orbital period of a planet round the Sun, or the orbital periods of any of the four bright satellites of Jupiter; any of these occurrences would be suitable regulators for measuring time, astronomically speaking. If a number of different clocks show various times at a given moment, one might well wonder which, if any, is right. Furthermore, if, on comparing a number of clocks with one which is used as a standard, it turns out that on average these clocks are all either in advance or all behind the standard clock by roughly the same amount, then it might be as well to question the accuracy of the regulator, or standard clock.

This is exactly what astronomers discovered at the beginning of this century, when they compared the various cosmic clocks. It was already known that the apparent acceleration of the Moon in its orbit round the Earth—a fact which conventional lunar theory was not able to account for—could be attributed to 'slow running' on the part of the Earth clock itself. Solar observations showed an apparent increase in the Earth's orbital velocity, and similar effects could be detected in the motions of the planets Mercury and Venus. Thus it became clear that the fault must lie with the timing device on which astronomers were relying, that the rotation of the Earth on its axis is not uniform, and that this is the real reason for the apparent disparity between practice and theory.

Brown, who was Newcomb's successor in Washington, and who worked out the final details of the lunar theory, came to this conclusion in 1914; later, in 1928, the Dutch theorist Willem de Sitter, who was at that time engaged on a study of astronomical constants, demonstrated that the Earth clock was by no means uniformly fast or slow, but that it was subject to fairly sudden changes. Abrupt changes could be shown to have taken place in 1897 and again in 1918.

It was at this moment that a technological discovery made it possible to tackle the problem of time without recourse to astronomical observations. In 1929, Harrison in Britain made the first successful quartz clock; this device uses the fact that a quartz crystal vibrates at a definite and constant frequency, when stimulated by an alternating electric field. The frequency of the vibrations in the crystal depends solely on the characteristics of its internal structure, and will therefore remain constant so long as these characteristics themselves are not changed as a result of some external stimulus (i.e. fluctuating temperature). The quartz clock provided astronomers with a method of measuring time which was well-nigh a hundred times more accurate than the conventional pendulum clock.

Continual checks on astronomical time determinations by means of quartz clocks substantiated the corrections required by the Earth clock, which apparent discrepancies in the motions of the Sun, Moon, Mercury and Venus, and the satellites of

Jupiter had indicated. Since 1935 it has been known that the Earth does not rotate uniformly on its axis; consequently, although it serves perfectly adequately as a timing device for ordinary purposes, it is unsuitable for work demanding extreme accuracy.

This would seem an apt place for some general remarks on time and its measurement. It is possible to measure the passage of time only in terms of certain definite intervals or pulses. We can express the time at which a given event occurred in multiples of certain agreed units—seconds, days, years—as reckoned from a particular, but nevertheless arbitrary, zero point. The various components of a clock form a vibratory system by means of which the unit of time is established, and the face and hands act as counters to register the number of time units which have passed.

The actual clock, in comparison with the idealistic clock (which would indicate the even passage of time), possesses at any given moment both a definite 'correction' as well as a 'rate'. The correction indicates to what extent the hands show the wrong time at that particular moment, while the rate indicates by how much the vibrations deviate from the theoretical value, that is to say the error compared with a uniform passage of time. Normally we do not differentiate between the two terms. If someone says 'My watch is fast', it can mean simply that the hands show a time which is in advance of what is true at that moment, or that the watch gains a certain amount during any given period. In the first instance the reference is to correction, and the error can be remedied by altering the positions of the hands, whereas in the second it refers to the rate which could be brought to zero by altering the vibratory mechanism within the watch.

In the case of the Earth clock, the hand is the direction towards the vernal point, while the celestial equator with its twenty-four hourly divisions represents the dial. In this 'clock' the dial moves past a mark. Astronomical time based on the Earth clock is liable to be erratic for two reasons, uneven movement of the dial, that is to say variations in the rotation period

45

of the Earth, and also the fact that the supposedly fixed mark, the vernal point, shifts.

This leads to the following thought: the effects of changes in the movement of a clock are cumulative, and over a long peroid even a slight error can build up to quite a considerable value. If, for example, the rotation velocity of the Earth reduces by as little as one ten-millionth, which would increase the length of the day by less than a hundredth of a second, then after one year astronomical time will be in error by something in excess of three seconds; in a century the error will have grown to a full five minutes. In celestial mechanics an error in time leads to faults in coordinating position and time; an error of one-tenth of a second of time results in an apparent displacement of position amounting to $1\frac{1}{2}$ seconds of arc.

It will have become obvious why a study of the unevenness of the Earth's rotation period is vitally important for celestial mechanics. If we examine the average length of the mean solar day during the period from 1750 to 1917 in terms of astronomical time, we find that the following corrections have to be made to the Earth clock: from minus 12 seconds in 1710 the value changed to plus 8 seconds by 1785; from that year until 1865 the value dropped again with some minor fluctuations to plus 3 seconds; in 1865 there was evidently a sudden change in the rate of the Earth clock, for the correction value sank to minus 8 seconds in the course of a few years (1880), and remained steady at this level until the next sudden leap in 1897. Then the rotation of the Earth began to speed up once more, shortening the length of the day by almost two milliseconds, so that by 1917 the time error had risen to as much as 20 seconds. As from 1917 the loss each day has been only half as much, but even so the value of correction had reached 33 seconds by 1960.

The realisation that the rotation period of the Earth is by no means a reliable time-keeper for astronomical purposes has led to a change in defining the unit of time. The new definition of a second of time is no longer founded on the mean solar day, the period of the Earth's rotation on its axis, and the tropical year is used instead; this is the interval between two successive passages of the Sun through the vernal point. Thus in terms of the new definition a second of time represents a 31,556,925·975th part of the tropical year 1900.

Coincidental with the new definition of the unit of time, a new concept of time itself was introduced: Newton's theoretical absolute time, which had temporarily been called mathematical or inertial time, was henceforth to be known as ephemeris time.

The same sort of task, which was entailed in reducing apparent positions based on the empirical system of coordinates to the inertial system of coordinates, now exists in converting empirical time to ephemeris time. The factors for this conversion have to be deduced from observations of the movements of bodies in the planetary system. This means that time corrections can be given only in retrospect, after observations over a given period of time have been evaluated and discussed, so long as we have to rely on clocks in which the measurement of time is not made independent of the processes of mechanical movement.

With the quartz clock we are already approaching such a new type of time measuring instrument. Its accuracy and freedom from the effects of external stimuli are not quite such as to provide the required uniformity of the passage of time over a decade or even a century which research now demands. Such high demands are likely to be met only with the development of clocks which rely on the vibrations of molecules (ammonia clock) or of atoms (cæsium clock). Once they have been set going, such timing devices would neither gain nor lose more than something in the order of one second in three hundred years.*

The efforts made and results obtained in the first half of the present century in order to determine and define the system of coordinates can be summarised as follows:

1. The system of coordinates represented by FK4 provides, with a very high degree of approximation, an inertial system.

2. As a result of observations of the Moon and the planets,

---

*Since 1967 the unit of time is based on the frequency of the Cæsium clock.

corrections to astronomical time based on the axial rotation of
the Earth give a fairly close approximation of ephemeris time,
which corresponds with Newton's concept of absolute time.

3. In the motions of the four inner planets—Mercury, Venus,
Earth and Mars—and of the Moon there are no longer any
irreconcilable discrepancies between observation and theory,
when for the purposes of computing the relevant perturbation
corrections based on the theory of relativity are taken into
account.

This last brings us to a point which has not yet been discussed. So far the emphasis has been on the classical fields, and
no mention has been made of the now famous words with which
H. Minkowski, a Göttingen mathematician, introduced his
lecture on 'Space and Time' at a congress of 'Gesellschaft
Deutscher Naturforscher und Ärzte' (Cologne 1908).

'The ideas of space and time which I would like to expound
have their roots in the field of experimental physics. This is their
strong-point. They tend to be radical. From now on space as
such and time as such will cease to exist as separate entities, and
only a union of both will exist.'

From that time on, no astronomer could allow a three dimensional system of spatial coordinates, a system which excluded
Einstein's fourth coordinate, time. All celestial bodies move in a
four dimensional Space–Time universe. Within the scope of the
discussion so far, the relativity effect has played no more than a
subordinate rôle, simply minor corrections to classical theory.
The significance increases when we are dealing with distances of
galactic proportions and with velocities comparable with that
of light.

4 · Structure and Dynamics of the Milky Way System

In the year 1915 a series of articles began in the *Astrophysical Journal* under the title 'Studies Based on the Colours and Magnitudes of Stellar Clusters'. The author of these studies was a young American astronomer named Harlow Shapley who was engaged in photographic-photometric investigations of star clusters with the 60-inch reflector on Mt. Wilson. He devoted his attention principally to what are known as Globular Clusters, a good example of which is that in the constellation Hercules. Article VII of the series (1918) bears the title *The Distances, Distribution in Space, and Dimensions of 69 Globular Clusters*; number XII, *Remarks on the Arrangement of the Sidereal Universe*. In these remarks the star system of the classical period, a concept first propounded by William Herschel on the basis of his star counts, and which Seeliger and Kapteyn thought they had described conclusively, was now superseded by a new idea, which envisaged a galactic system from five to ten times greater in its dimensions, and in which the Sun was sited near the periphery instead of at the centre. A sort of new 'Copernican' turning point was thus established, having been brought about through the application of improved methods of distance determining, methods which no longer relied solely on geometric principles—light rays are straight lines—but on the physical nature of stars themselves and on the belief that the laws of physics were universal.

During the early twenties there was a great deal of discussion concerning this 'Greater Galactic System' and Shapley's proposed scale of distances. There were also attempts to reconcile the old with new, the idea of a subordinate system; descriptions such as 'Local System' or 'Local Cluster' were coined to meet the situation. In the minutes of the Bavarian Academy of Sciences in 1920 Seeliger described a 'Typical Star System' as he called it, based on the most recent star counts then available. Kapteyn did likewise in 1921 in the *Astrophysical Journal* under the title 'First attempt at a theory of the arrangement and motion of the sidereal system'.

Alternatively efforts were made to try to reduce the dimensions of Shapley's Greater Galactic System. Charlier suggested a view of the system of globular clusters in which their apparent diameters are used to establish a scale of distances. Pannekoek's method of determining distances was based on the distribution function of luminosities. Seeliger attempted to show that, through absorption of light, Shapley's scale of distances was markedly in error. All these efforts succeeded in shaking the foundations of the photometric method of distance determining, so that the main task in the twenties and thirties was the clarification of this point.

The uncertainty of the situation was emphasised in a report by August Kopff in Vol. II of *Ergebnisse der exacten Naturwissenschaften* (1925); he writes: 'It seems invariably the fate of the typical system to appear inadequate so soon as the transition to the actual system is made. The typical system must be adapted so as to correspond to reality. However, the proposed distances far exceed admissible limits. . . . Contradictions really do exist, and their solution is not yet clear. Perhaps, in the process of establishing a mean trend in star counts, stellar voids and accumulations just happen to have occurred and so we have the impression of a gradual reduction in stellar densities as we move outwards. . . . It could also be the case that the distances proposed by Shapley are vastly in excess of their true values. . . . But we could only reasonably assume a lower distance value in the universe, if the relationship between absolute luminosity and spectral properties, or else the behaviour of variables as we know them in the neighbourhood of the Sun, do not apply universally. This is not very probable; observational information to date favours considerable physical homogeneity in the universe.'

In an obituary of Seeliger in 1924 I wrote: 'Seeliger's fundamental works, like those of Kapteyn, so far as they are based on star counts and mean parallaxes, are a part of history. Opinions have changed. The time has come when the fiction of a universally applicable distribution function of absolute luminosities must disappear from arguments concerning the structure of the universe. The permeation of astronomy by physical theories means that a particular star's spectral classification is of greater significance than its absolute brightness, and in any case precludes independent consideration of either. Seeliger's Typical System may be summarised as follows: At the centre, families of stars passing through each other, which betray their common origin in traces of a common motion and similarities of state; at the edges the system merges more or less gradually into individual Milky Way clouds, and the whole appears to be embedded in one large system whose members are globular clusters and spiral nebulæ.'

In 1926 Gunnar Malmquist, a young Swedish astronomer of the school of Charlier, reported on an investigation in which he took a model star system of given density distribution, and calculated the number of stars theoretically visible from the centre of this system; then, on the basis of his results, he used the methods of classical stellar statistics to derive the density law from these calculated star numbers, that is to say the spatial distribution of stars in the system.

The model he chose had a very simple characteristic density distribution: a sphere uniformly filled with stars surrounded by a shell also uniformly filled with stars, and between the two a void space of the same thickness as the outer shell.

Analysis of the star counts gave a picture which was quite different from actuality. The density of stars appeared to fall off uniformly from the centre outwards, rather like a similar sort of law which Seeliger and Kapteyn had evolved for the Milky Way system. The starless zone of the real model was not at all evident —a truly alarming result, from which it was necessary to deduce that perhaps the old star system had no real substance. On the basis of what we now know about the structure of the stellar system it is not difficult to see why the methods of stellar statistics, which rested solely upon counts of stars according to their apparent magnitudes and proper motions, inevitably led to a false impression.

In the first place, the simplifications introduced for a formal mathematical treatment of the problem were too sweeping. The system cannot, even in the first approximation, be described by means of a density function dependent on distance from the Sun and vertical distance from the plane of the Milky Way (which is assumed to be the plane of symmetry), and a distribution function of absolute luminosities independent of location. Through the smoothing of observed numbers of stars, in particular the blurring of distinctly cloud-like irregularities in the distribution of stars along the band of the Milky Way, the characteristic trends of actual spatial distribution are lost. The direction towards the true centre of the system, indicated through accumulations of very faint stars and globular star clusters, and lying far from our Sun, could not emerge from the formulæ from which the density distribution was deduced.

Furthermore, the empirical data of star counts, which were available for such calculations, and the mean parallaxes derived from apparent proper motions were not sufficiently comprehensive to include such faint stars; consequently they could hardly be expected to reveal conditions in the more distant reaches. Evaluation of the data gained by Kapteyn as a result of the plan of 'selected areas' brought no change in this state of affairs. It soon became evident that the star fields, which had been selected in the light of the classical concept of what constituted a typical system, were not representative of the real Greater Galactic system.

Yet another consideration is probably the underestimation of the effect of absorption of light as a result of diffuse matter scattered through interstellar space, clouds of gas and cosmic dust, which manifest themselves as starless patches.

As from about the year 1924 no doubt remained regarding the reality and the dimensions of the Greater Galactic System as defined through the globular clusters. In a discussion at the National Academy of Sciences in Washington in April 1920 between Harlow Shapley and H. D. Curtiss on 'The Scale of the Universe' the arguments in favour of the greater system eventually proved themselves more convincing; the idea of physical unity in the cosmos had become established. Further investigations could only fill in the details and determine an exact scale of distances, that is to say the distance of the Sun from the centre of the system. Hence the methods developed for stellar statistics gave way to the study of individual objects, and could be used only in a limited sense for groups of objects, which had been selected according to various sorts of physical parameters. Spectral class, absolute magnitude, type and period of variability of Cepheid stars, and the velocity and direction of spatial movement are the most important of these parameters.

With refinements in the analyses of proper motion and radial velocities a new insight was gained. It was soon realised that the direction of the apparent movement of the Sun differed according to the magnitude of the particular stars on whose proper motion the calculations had been based. Similarly, when calculating the velocity of the solar motion from the radial velocities of stars, it was found that the value differed with the spectral class, and that there was a marked difference between giants and dwarfs, and this even led to the discovery of a group of stars with especially high velocities.

By about 1924 the various findings had coalesced into the following views:

1. The velocity of the Sun is in the order of 20 km./sec. relative to the stars in its immediate vicinity, and 35–50 km./sec. relative to those lying at a greater distance, while it exceeds 200 km./sec. relative to the system of globular clusters.

2. The direction of apparent solar motion relative to the system of globular clusters is almost perpendicular to the direction towards the centre of the galactic system.

3. The directions in which the high-velocity stars move fall into two distinct groups symmetrical with the direction towards the centre.

The significance of this phenomenon was explained simultaneously by Lindblad in Uppsala, and by Oort in Leyden in 1926–27, namely that the Milky Way system is in rotation about the centre determined by Shapley. The high velocity observed in the Sun's movement relative to the globular clusters is due to the Sun's orbital velocity. The high-velocity stars are stars travelling in eccentric ellipses rather similar to those of the comets in the planetary system.

The picture presented by the motions of the stars in the immediate neighbourhood of the Sun is essentially determined by the fact that the star system does not rotate as a rigid body. In fact orbital velocity tends to decrease with distance from the centre of the system; those members of the system which lie closer to its centre than the Sun race ahead of the Sun, while those lying farther out are left behind. This differential rotation can be detected in the characteristics of proper motion and radial velocities relative to the Sun, if one plots the values of the observed velocities against their direction. In four directions at right angles to each other they are at zero; in the intermediate directions they attain alternate positive or negative maximal values.

This 'double wave' in the radial velocities is one of the main props of the rotation theory. From it the constants of galactic rotation may be calculated—that is to say the direction towards the centre of rotation, how far this centre is from the Sun, and the orbital velocity of the Sun around this centre. The result of Oort's first calculations showed the centre of rotation to be identical with Shapley's centre of the system of globular clusters. The Sun lies about 10,000 parsecs from the centre of the system, and travels at an orbital velocity of approximately 250 km./sec., so that the orbital period is something in the order of 250 million years.

In developing further details of the rotation theory it had to be borne in mind that movement within the Milky Way system is not, as in the Solar System, governed by a body of over-whelmingly large mass; consequently the orbits of the stars are by no means simple Keplerian ellipses, let alone circles. The decisive factor governing movement is to be found in the spatial distribution of masses. Since not all these masses are directly visible as stars, or even luminous gas clouds, a great deal depends on determining the mass and distribution of interstellar matter.

Research into interstellar absorption thus became doubly important: first, because light is weakened in its passage through the interstellar medium and thus falsifies the results of photo-telemetry; second, because the light is not only weakened, but also changes its colour—the greater the distance through which the light from a star has to travel the redder it appears. This pro-vides a clue to the composition and mean density of interstellar matter.

The sort of ideas which had grown up by about 1930 con-cerning the structure of the galactic system can be learned from some short extracts from *A Textbook of Astronomy* by Elis and Bengt Strömgren (father and son), the foreword to which is dated 1932:

'Open star clusters, globular clusters, loose clusters, Cepheids, red long-period variables, and novæ create a spatial impression of the following nature: two great systems at some distance from each other; the globular clusters primarily belong to one system, though some lie far from its centre. These two prin-cipal systems are usually referred to as the local system and the Sagittarian system.

'It is quite possible, even probable, that the system, which we call local, extends along the plane of the Milky Way. There is therefore a possibility that we belong to a coherent Milky Way system, which is arranged along one and the same plane with the Sagittarian system as the centre of the whole. The diameter of the system along the plane of the Milky Way would then be in the nature of 30,000 parsecs, and the Sun is located fairly eccentricly.

'On the other hand, there is also a possibility that the Sagittarian system and the local system are both limited in their extent, which would mean that the Milky Way system is made up of these two systems, and a few globular star clusters (lying far from the Sagittarian centre), as well as perhaps some other system or systems as yet unknown.

'The idea of the entire Milky Way system as one coherent system having rotational symmetry about an axis passing through the Sagittarian centre and perpendicular to the plane of the Milky Way has in its favour the fact that it allows a single explanation for the motions of bodies in the vicinity of the Sun. Against this there is the argument that, apart from the Milky Way system, we know of no external system whose diameter is in the order of 30,000 parsecs, although super-systems consisting of a number of smaller individual systems, and having the required overall dimensions have been observed.'

In the years that followed there was a trend towards reconciling dimensional differences between our Galaxy on the one hand, and the Andromeda Galaxy on the other. Taking the absorption factor into account in determining the distances of the globular clusters, brought about a reduction in the dimensions of the Milky Way system, above all its diameter. By means of direct photo-electric measurements of the Andromeda Galaxy it became possible to trace the turns of the spiral arms much farther outwards than the photographic method had allowed; this suggested that the Andromeda system is in fact larger than Hubble had supposed.

Comparable studies of the Andromeda Spiral on the one hand and the Milky Way system on the other began to point more and more definitely to the conclusion that the two systems are essentially similar. We can view the Andromeda system as a whole from our distant vantage point in space; consequently many of its features appear simplified and more clear-cut. From 1935 at the latest, one could consider the alternative (i.e. the Milky Way as a form of super-system) as having been amply refuted, and one was in a position to say: 'The flat, lens-shaped system of the classical epoch of stellar astronomy is a myth. It is not possible to reconcile it with the Greater Galactic system as defined by the globular clusters simply by enlarging the scale, nor can it be included in such a system as a distinct subordinate system. The Greater Galactic system is an entity; the globular clusters form its frame-work; the Sun is situated 8,000–10,000 parsecs from the centre, in a star cloud within one of the spiral arms.'

The work of one particular man has been of immense significance so far as the clarification and confirmation of our understanding of the structure of the Milky Way system is concerned. He knew just how to get the best out of the two largest instruments available, the 100-inch on Mt. Wilson and the 200-inch on Mt. Palomar. Walter Baade was born in 1893, and after graduating at Göttingen became assistant to Felix Klein, the mathematician; he first made a name for himself through his work with the 1-metre reflector at the Hamburg Observatory. At the Mt. Wilson Observatory, where he became one of the permanent assistants, he was soon acknowledged as an authority on problems relating to the structure of the Milky Way system and the extragalactic universe.

In 1943, using the 100-inch, Baade succeeded in achieving partial resolution of stars in some of the elliptical nebulæ, as well as in the inner regions of the Andromeda Galaxy. Success came not through sheer chance, but through the idea that the brightest stars of the extragalactic systems were likely to be red giants, just as the brightest stars of the globular clusters, and should therefore register individually on panchromatic photographic plates, if a strong red filter were used to cut out the background of weaker non-resolvable stars. The success of this method not only brought confirmation of Baade's theory concerning the existence of two different star populations, but also its practical application in the study of the Milky Way system. The various stellar spectral types occur in a different ratio at the core of star systems than in the spiral arms of the outer regions; in Population I of the spiral arms bright blue or white stars predominate, while in Population II of the nucleus there is a predominance of red giants.

A few years later at a conference in Michigan on 'The Structure of the Milky Way System', Baade was able to announce further results of his investigations of the Andromeda Spiral. He found that there were also stars of Population II in the outer regions, and that they had even been detected by photo-electric methods in the outermost limits of the system, whereas stars belonging to Population I were confined to the spiral arms, where they were associated with emission nebulæ and interstellar matter.

The inference drawn from this knowledge about the Andromeda Galaxy on the structure of the Milky Way system found surprising corroboration in 1951 as a result of certain efforts in the field of radio astronomy. The sources of 21 cm. radiations emanating from interstellar hydrogen trace the extent of at least three distinctly defined spiral arms, one of which contains the Sun.

At the same time, at the International Congress in Rome, Baade made the exciting announcement that the distance and thus also the dimensions of the Andromeda Spiral as calculated by Hubble should be doubled. In this way the last remaining discrepancy, which had been such an embarrassment in the early 1930s, the fact that the Milky Way system was apparently so much bigger than the extragalactic systems, was swept away. The Milky Way system is well within the observed range of diameters of galaxies, and does not differ essentially from the Andromeda system.

The concept of the Milky Way system, which replaced that proposed by Karl Schwarzschild in 1909, may be summarised as follows: The Milky Way system is similar in structure to the great spiral galaxies with their clearly recognisable arms. There is a distinction between the nucleus, a sort of extra large globular accumulation of stars which has a diameter of about 15,000 light-years, and the surrounding flat disc containing the spiral arms which has a maximum extent of 100,000 light-years in the main plane of the system, and a thickness of only about 3,000 light-years; around the edge of this is a sort of globe-shaped halo, whose diameter is at least 150,000 light-years, and which consists of a loose scattering of stars and globular star clusters.

This remarkable, and to a certain extent geometric, division into three distinct parts, corresponds to differences in the distribution of stars, star clusters and interstellar matter. Globular clusters, long-period variables, and high-velocity stars, now thought to be the oldest members of the system, make up the population of the halo. In the disc we find not only stars of the same kind as those which occur in the nucleus, but also planetary nebulæ, short-period variables, and stars of manifestly low metallic content, all of them also relatively ancient objects. Along the spiral arms bedded in this disc are the stars of Population I, the galactic (open) star clusters interspersed at random with interstellar gas and dust. The population of the spiral arms contains the youngest members of the system.

The total mass of the entire system is something in the nature of 100,000 million to 200,000 million times that of the Sun. About 23% of the mass is evenly distributed through the halo; the disc and nucleus account for 67%, while the remaining 10% represent the content of the spiral arms. Ninety per cent of the overall mass is concentrated in stars; 10% is in the form of diffuse gas, while the dust content is no more than something in the order of one per thousand. The position of the Sun within this system can be given by means of the following references: distance from centre 25,000 light-years, height above the galactic plane north 45 light-years.

The rotation is like that of the Andromeda Spiral; the innermost part (nucleus) turns at constant angular velocity, that is to say it behaves like a solid body; outwards from the nucleus, angular velocity decreases with distance. In the region where the Sun is located rotational velocity is about 217 km./sec., equivalent to an orbital period of 234 million years or so. Some of the stellar orbits take the form of very eccentric ellipses which bring stars from the region of the nucleus out to where the Sun lies. None of the orbits, however, seems to be more than slightly inclined to the general galactic plane.

5 · The Realm of the Nebulæ

'What are galaxies? No one knew before 1900. Very few people knew in 1920. All astronomers knew after 1924.' Words to this effect appear in the introduction to the *Hubble Atlas of Nebulæ*. It then goes on to explain: 'Galaxies are the largest single aggregates of stars in the universe. Each galaxy is a stellar system somewhat like our Milky Way, and isolated from its neighbours by near empty space. In popular terms, each galaxy is a separate "island universe" unto itself.'

The dilemma, which the debate between Shapley and Curtiss before the National Academy of Sciences (Washington, 1920) emphasised, was due to the apparent inability to reconcile the scale of Shapley's proposed Greater Galactic system, and the idea that this system and the Andromeda Spiral were in any way comparable in their dimensions. A decision either for or against the island nature of the spirals could be given only when it became possible to determine the distances of these objects—much the same as in the case of the globular star clusters.

The principle which is used for these measurements lies in the connection between the period of the light change of a particular type of variable star, and luminosity; this is known as the 'Period-Luminosity Law' of the Cepheids. Cepheids in other systems were first discovered in 1922 (Messier 33), and in 1923 (New General Catalogue 6822); towards the end of 1923 Hubble was able to obtain the positive identification of the first Cepheid variable in the Andromeda Spiral.

An observation programme with the 100-inch telescope which was begun at this time rapidly led to further discoveries. By the end of 1924, already thirty-six variables had been detected in the Andromeda Spiral, twelve of which were recognised as Cepheids; in M.33 the number was twenty-two Cepheids, and in NGC 6822 eleven. This provided proof of the extragalactic nature of the spirals, and also established a scale of distances for the galactic universe. Thus, for the Solar System we have the Astronomical Unit (AU = 150 million km.), and within our Galaxy, the Milky Way system, the unit is the 'kiloparsec' (= 206 million AU = 3,260 light-years), but when we are dealing with distances between the galaxies even these units are too small, and we must reckon in 'megaparsecs' (= 206,000 million AU = $3\frac{1}{4}$ million light-years!).

The Swedish astronomer Knut Lundmark made an important contribution to this question of distance. He was the first to estimate distances of the order of several million light-years based on the appearances of novæ in the spiral nebulæ, working on the assumption that these novæ were likely to be of the same kind as those we know in our own galaxy, and that these phenomena share a common physical background. He suggested that 'all novæ attain at their peak the same absolute brilliance'. This hypothesis has since been corroborated but with the modification that there are in fact two kinds of these stars: the ordinary novæ, and the supernovæ, whose luminous intensities differ by a factor of 100 to 1,000.

From 1925 to 1936 there appeared in the *Astrophysical Journal* a series of articles under the heading 'Studies of Extragalactic Nebulæ', just as in the previous decade there had been articles by Harlow Shapley on the globular clusters and the Greater Galactic system. The author of this series was Edwin B. Hubble; his instrument was the Hooker Telescope, the 100-inch reflector of the Mt. Wilson Observatory, which in 1921–22 yielded the first photographs of NGC 6822 and M.33 revealing star-like condensations, and subsequently led to the discovery of some variables. In his book *The Realm of the Nebulæ*, published in 1936, Hubble described all that was then known about them.

61

Here are some of the more pertinent points from the book which will show the sort of ideas current during that decade.

The first fixing of a scale of distances presupposes the resolution of the nearest objects, and the identification of individual star-like features on photographs as true stars. If the distance of the Andromeda Spiral is something in the order of a few hundred thousand light-years, then even the smallest distinguishable and apparently single point of light could nevertheless be a collection of stars in actuality; this is because at a distance of 500,000 light-years any group of stars which occupies a space one hundred times as large as the Solar System (diameter 6,000 AU = 0·1 light-years) appears no bigger than 0·04 seconds of arc, and, even under the best possible observing conditions, this is beyond the resolving power of a 100-inch instrument, since in such a telescope the theoretical image of a single star has a diameter of this order.

It is possible to decide whether a particular object happens to be a single star or a group of stars, if the star-like condensation displays a variation in its brilliance similar to that of certain variables in the Milky Way system. Since the light curves worked out by Hubble from numerous photographs of the Andromeda Spiral taken over a long period of time completely agree in form, period and amplitude with those of classical Cepheids in our Galaxy and in the globular clusters, it seemed fairly certain by the year 1924 that these smallest elements of resolution were in fact individual stars of great intensity.

The next step was to compare other bright star-like objects with the bluish and whitish stars, as well as the reddish irregular variables in the Milky Way system, thus making it possible to determine the distances of galaxies still farther away; for in these distant objects only the brightest stars can be distinguished from the general background haze of fainter stars, and the Cepheids are utterly swamped. Having established the range of the Andromeda Spiral at approximately 900,000 light-years, the insular nature of the galaxies was adequately proven, but the general similarity with the Milky Way system was by no means clear as yet. There was a discrepancy which could be stated in the form that, 'if nebulæ were island universes, the galactic system was a continent' (Hubble).

It has already been explained how, in the course of time, new discoveries helped to resolve this discrepancy. The 'continent' shrank as a result of revisions of the yardstick with which it had been measured, while the 'island' grew through the discovery of its outermost contours. Some disagreement remained in a comparison of individual features found in both systems. The nebulous-looking objects found by Hubble in the Andromeda Spiral in 1932, and provisionally identified by him as 'globular clusters', were apparently one to two times fainter than corresponding clusters in the Milky Way system.

In an intensive search of the heavens for extragalactic objects, whose numbers soon swelled to millions, it became evident that their apparent distribution in the sky, as well as their actual distribution in space, was by no means uniform, but that instead galaxies often tend to occur in groups. Already at the beginning of the century, Max Wolf had suggested the existence of such 'nests of nebulæ', and discovered a great cluster of nebulæ in the constellation of Coma Berenices in 1901, and in 1905 that in Perseus. Then in the twenties Shapley and Miss Ames carried out their investigations of the heavens at the Harvard Observatory, and this served as a basis for the study of the spatial distribution of the nebulæ.

There emerged from these investigations one particular group or cluster which subsequently became known as the 'Local Group', and held a special significance because our own Milky Way system seemed to belong to it. In his list of 1936, apart from the two great spirals M.31 in Andromeda and M.33 in Triangulum, Hubble included as members of this Local Group the two elliptical companions of the Andromeda Spiral and four 'irregular' systems of the same type as the Large and Small Magellanic Clouds of the southern hemisphere. The diameters of these systems range from 800 light-years for M.32, the almost spherical companion of the Andromeda Spiral, and 40,000 light-years for the Andromeda Spiral itself, while the combined luminosity is equivalent to between 2 million Suns on the one hand and 1,000 million on the other. This wide range both in dimensions and in luminosities has to be taken into account if the distance of a given system is to be deduced from its apparent diameter and magnitude, when it is too far away to be resolved into individual stars, and can only be observed as a whole. Thus a 'luminosity function' plays a part, in the same way as in the methods of classical stellar statistics, for a study of the structure of the Milky Way system.

Research into the movements of spiral galaxies brought unexpected, and at first incomprehensible, results. In 1914 Slipher, at the Lowell Observatory in Flagstaff, Arizona, published his first list of fifteen radial velocities, which were used to calculate the movement of the Sun relative to the spirals. The velocities were found to be not only extremely high (between 300 and 1,800 km./sec.) in comparison with those of stars, but also, with very few exceptions, directed away from the Sun. The idea of a general dispersal of the galaxies at a mean velocity in the order of 600 km./sec. became increasingly obvious as observational data accumulated, and at the same time it could be seen that there was a connection between velocity and distance. In 1929 the radial velocities of forty-six spiral galaxies had been listed, and for twenty-four of these the distances were also known. They covered a range of distances of 6 million light-years, and radial velocities up to 1,100 km./sec.; for every million light-years velocity increased by about 170 km./sec. This is the 'velocity–distance relation' which Hubble reported to the National Academy of Sciences in March 1929, and which is now generally known as the 'Hubble Relation'.

If this relation expresses a universally applicable law governing the velocity distribution of the galaxies, analogous to Kepler's Third Law which shows the relationship between orbital periods and distances of the planets from the Sun in the Solar System, then we may employ it as a means of determining the distances of faint galaxies, where photometric methods, which must rely on the ability to resolve the brightest stars, fail because this is no longer feasible. In order to plumb the farthest extent of the realm of the nebulæ, work began at the Mt. Wilson

Observatory with the 100-inch to try to determine the radial velocities of the very faintest systems which that instrument could pick up. By 1926 the first piece of irrefutable evidence was available, a velocity of almost 4,000 km./sec., which would therefore correspond to a distance of some 25 million light-years.

The inferences drawn from Einstein's General Theory of Relativity concerning the possible structure of a space–time universe allowed the relationship between velocity and distance to be seen in another light. At the end of his report Hubble wrote as follows: 'Meanwhile it is worth noting that the velocity–distance relation might perhaps be an example of the de Sitter Effect, and that, as a result, the numerical data could enter into discussions on the general curvature of space.'

In 1916–17 de Sitter had found a solution of the equations of general relativity which led to an open hyperbolic world model in contrast to the closed spherical model which Einstein had derived. In the de Sitter universe, a red shift should be observable in the spectra of the galaxies, and, if this is interpreted as a Doppler Effect, would lead to the general impression of outward-fleeing nebulæ.

Its importance regarding a solution of cosmological problems gave the Hubble Relation a prominent place in debate during the years that followed. After the de Sitter universe had been recognised as a special instance of a non-static universe, and a whole multitude of theoretically possible models for non-static universes had been formulated, it seemed for a while as if the Hubble Relation might supply the answer as to which of the models possible in theory was likely to be most representative of the real universe. Before this could happen, however, observations had to include velocities more or less comparable with that of light, because it is only in such circumstances that the inference of the theory becomes at all clear. At the same time, the distances of objects with such high radial velocities would have to be determined through independent methods— from apparent magnitudes or diameters—if any deviation from linearity were to be detected in the relation.

Considerations regarding the age of the universe are closely coupled with those of an expanding universe. If what we are seeing today is the universe in a state of expansion, then this fact immediately raises the problem of what might have caused such expansion to take place. Further, if one imagines the phenomenon to be similar to an exploding grenade, it raises the question at which point of time in the past the explosion of this 'grenade' might have taken place. If it is assumed that the Hubble Relation expresses the rule according to which the expansion is proceeding, then it is possible to calculate from this how much time has elapsed since the universe began to expand. Supposing that the expansion is taking place uniformly, it follows from the Hubble constant of 170 km./sec. per million light-years, which at that time was still thought to be correct, that the universe as such was barely 2,000 million years old.

This then was the state of affairs around the year 1935. Fifteen years later, Walter Baade reported on *Galaxies—Present Day Problems* and in his writings outlined the developments which had taken place. 'In the 25 years which have elapsed since the extragalactic nature of the spiral nebulæ was established the centre of interest in this field of research has shifted in a most remarkable manner. It all started normally enough. Through the discovery of Cepheids in these systems their stellar composition became firmly established, and subsequent investigations showed that like our own galaxy, star clusters, gas, and dust are mixed with the stars of these systems.

'But hardly had the first provisional data for a few of the nearest galaxies been obtained, when a great discovery, the red shift in the spectra of the nebulæ, diverted the inquiry into new directions. You all know what followed. The cosmological problem became the dominant question and a tremendous effort was made towards its solution.

'We know today that this bold first attempt ended in failure. The reasons of this failure are in part of a technical nature, as for instance the provisional character of the photometric scales used in this first attempt. There is no doubt that with the means now at our disposal reliable photometric scales will be available

66

in the near future, although the requirements demanded by theory in order to distinguish between possible types of universes remain uncomfortably stringent. We are not so certain any more, however, whether certain basic assumptions underlying this first attack are justified. . . . Altogether, there are good reasons to believe that solution of the cosmological problem is much more difficult than was thought some 15 years ago and that it may well lie beyond our present powers. Certainly, we will need much wider and better-secured foundations than at present before we can hope to erect big superstructures.'

The possibilities of the 100-inch telescope at Mt. Wilson had by now been exhausted, and any further advance depended on the new 200-inch Hale Telescope under construction at the Mt. Palomar Observatory; in particular the definition of a new photometric scale for the faintest galaxies based on photo-electric measurements, and the measurement of red shift in the spectra of the weakest and farthest removed galaxies receding from us at velocities approaching that of light itself.

The first revision of Hubble's distance scale, which Baade announced at the Astronomical Congress in Rome in 1952, had a disturbing effect to start with, since it queried one of the basic assumptions, namely the existence of a single and universally valid 'Period-Luminosity Relation' for Cepheid variables. Baade's differentiation between two types of star populations meant that it was also necessary to differentiate between two kinds of variables, those of Population I and those of Population II. The short-period variables, so called RR Lyræ stars, which in the Milky Way system occur mainly in the globular clusters, have periods of less than one day and are typical representatives of Population II; on the other hand the long-period variables, classical examples of Cepheids, in the vicinity of the Sun belong to Population I. Classical Cepheids do also occur in systems with Population II stars, but they are on average approximately 1·5 magnitudes, that is to say four times as bright as the typical Cepheids of the same period of Population I.

The period–luminosity relation which Shapley used for determining the distances of the globular clusters applies to stars of Population II. The variables discovered in spiral galaxies by photometric means are, however, those which are four times brighter than the typical variety, and thus belong to Population I. Hence Shapley's period–luminosity relation gives the right answer in the case of globular clusters, that is to say the correct scale for the Milky Way system; in the case of extragalactic objects, however, distances computed on this basis are too small by a factor of two. This means that Hubble's scale of distances for the galaxies needed at least to be doubled.

In the course of further investigations, it has become apparent that a distinction between only two types of stellar populations is inadequate, and fails to explain all the phenomena which the stars and star systems display. Later we shall discuss how, from about 1940 on, the idea of the evolution of stars and star systems gained ground, and how the age of a star or system plays an important rôle in many considerations. At this stage, suffice it to say that Baade's differentiation between two kinds of star population on purely phenomenalistic grounds actually entails a continuous succession of stages according to age. In place of one period–luminosity curve there is really a corresponding succession of curves with various zeros. The luminosity of a variable is no longer determinable solely on the grounds of period; further criteria are needed which will indicate the age of the star concerned. In this way the method of determining distances through variables suffers a slight loss of reliability.The inclusion of other objects, as for instance novæ and supernovæ, or globular clusters, as well as the question of age, led eventually to a further revision of Hubble's distance scale.

The effect of this on astronomical thought is shown by the following examples of distances considered for the Andromeda Spiral:

1936	750,000 light-years
1950	1,500,000 light-years
1956	2,300,000 light-years

Apart from an alteration in the zero point of the period–luminosity relation, the revision of the photometric scale affects the velocity–distance relation insofar as the distances calculated from the apparent magnitudes of very faint galaxies are much too great. The numerical value of the Hubble constant, which shows the increase in the velocity of recession per megaparsec, was reduced from 530 km./sec. in 1936 to 75 km./sec. in 1958. Consequently the 'expansion age' of the universe is nowadays taken as 13,000 million years instead of the bare 2,000 million years as thought in 1936.

If we now turn our attention to the structure of the system of galaxies and try to discover the laws which govern this structure, then the tendency to collect in groups or clusters is an important characteristic of the apparent and spatial distribution; the number of systems to be found in comparative isolation in space is small.

On average about five galaxies occupy a space of approximately 100 cubic megaparsecs. In a model where the galaxies are represented by lens-shaped objects with a mean diameter of 5 mm., 400 such 'lenses' would have to be arranged in a space measuring $10 \times 10 \times 10 = 1,000$ cubic metres; however, they should not be arranged uniformly at a mean distance of about 2·5 metres from one another. In order to obtain the real picture, the objects would have to be divided into perhaps two or three separate groups, each no more than half a metre in diameter, so that within each group the 5-mm. lens-shaped objects would be about 25 mm. apart on average, while in the void between the groups there would be an isolated 'lens' here or there. In our earlier model of the Milky Way system, we spoke in terms of pinheads at distances in the order of 100 km. from each other, reducing to about 1 km. in the densest aggregations; but here the picture is wholly different.

If the galaxies are separated by distances which are comparable with their own dimensions, that is to say relatively close to one another, then one can count on considerable reciprocity between the systems, in contrast to conditions within the systems themselves, where the distance from one star to the next is of quite a different order of magnitude. Deformation of a galaxy in consequence of the tidal effects exerted by neighbouring systems can be such that portions are torn free, and streams of matter move from one system to another. Countless examples of such bridges of matter between galaxies are known, many of which have been discovered due to the work of Zwicky. It was also thought that there was evidence for collisions between galaxies, though it is doubtful whether the photographs in question really do show galaxies passing through each other.

The question of whether there exists a super system, whose elements are the galactic clusters, is one which still remains very wide open. There are certain indications in favour of the idea that the local group, together with several other galactic clusters, forms a part of a sort of super-galaxy which has the great Coma-Virgo cluster as a nucleus, and about which it rotates with a period ranging between 50,000 million and 150,000 million years.

Equally open is the question concerning the inner structure of the clusters of galaxies, which display considerable variety both in their sizes and in the numbers of component members. At present we differentiate between four kinds, dependent on the number of components, the degree of concentration, and the regularity of arrangement (spherically symmetric or cloud-like). It is possible that the different sorts represent different stages of evolution.

A final question which is still unresolved concerns the wide variety of sizes, masses and luminosities of the galaxies, and their connection with other cosmic structures, in particular the connection between the series of elliptical galaxies and the globular star clusters. In 1944 Baade emphasised that between the smallest galaxies with diameters ranging from 800 to 1,000 parsec, and the largest globular star clusters, whose diameters are only in the order of 100 parsecs, there is a wide gap which is unlikely to be filled. In the meantime, however, as a result of a systematic study of clusters of galaxies Zwicky has discovered some dwarf galaxies containing no more than some thousands of stars, and measuring down to 500 parsecs in diameter. Perhaps the gap, to which Baade referred, will be filled in the course of the next few years.

6 · The Hertzsprung–Russell Diagram

It is something of a precarious task to try to describe solely in words something that is a pictorial representation of physical relationships, without being able to drive home the point with an appropriate line. It also makes great demand on the imaginative powers of the reader, and I am not at all sure that I shall succeed in conjuring up the correct image. If the attempt proves successful, it will lead to a real understanding of the problems under discussion. I would suggest that the reader equip himself with pencil and paper, so that he may draw up the diagram from the description below.

Let us begin by enumerating the concepts with which we shall be dealing: magnitude, colour and spectrum.

We talk of stars of the first magnitude, and second magnitude, and so on; by this we mean their apparent magnitudes, i.e. how bright they appear to be. The apparent magnitude is by no means a direct physical characteristic, because it depends in part on how far from us that particular star happens to be. What we need to know in order to distinguish the real radiation characteristics of the stars is the *absolute* magnitude, the actual luminosity, which is independent of distance.

The measure for the luminosity of the stars is our own Sun; for astronomical purposes this acts as the standard candle with which the stars are compared. The luminosity of a star is in fact equivalent to the candle-power of a light bulb, only more so; the cosmic 'candle' is the Sun, equivalent to a 400 quadrillion-watt bulb. The range of stellar luminosities is about as great as that between a glow-worm and a million candle-power searchlight; in this comparison the Sun may be considered about as powerful as a fairly modest nightlight.

In contrast to the apparent magnitude, the colour of a star bears a direct relationship to a particular physical state, namely the surface temperature. With an increase in this temperature the colour changes from deep red through yellow into intense white, and in the case of very high temperatures assumes a distinctly bluish tinge. The 'colour index' of a star is the difference in its apparent magnitude in two differing wavelengths. With the help of the laws of radiation, it is possible to calculate the surface temperature of a star from the colour index.

The colour or colour index of a star represents only a small fraction of the wealth of information available in starlight; a detailed analysis of these radiations teaches us yet more. From two-colour photometry we can go on to three- or four-colour photometry so as to obtain a more complete picture of energy distribution in the form of a continuous spectrum, and to determine the intensities of the lines which overlie this spectrum.

The sequence of colours from blue through white and yellow to red denotes a series of spectral classes, which have been redefined and added to from time to time. The classification developed for the Henry Draper Catalogue of the Harvard Observatory is the version which was finally adopted all over the world. Each of the letters O B A F G K M is used to signify a certain basic type. The choice and sequence of these letters may seem rather haphazard, but they are the relic of a previous code in which classification was carried out according to the intensity of certain spectral lines, and identified with letters which originally ran in the normal alphabetical sequence. With increasing insight into physical occurrences the series was altered to denote a descending order of temperatures in equal steps. In this way some of the letters were left out altogether, while in other cases the alphabetical order had to be rearranged. In addition each of the letter classes consists of subdivisions numbered from 0 to 9.

The sequence can easily be remembered with the aid of a

72

mnemonic, which, so far as I know, originated with Henry N. Russell: '*O Be A Fine Girl; Kiss Me!*'—Sometimes three further classes bearing the letters R, N and S are attached to the list, so that the mnemonic ends with the words '. . . *Right Now, Sweetheart*' (though some say the last word should be '*Smack*').

I propose to avoid using these fairly exact definitions so far as I can, and to try to confine myself to wider terms such as blue, white, yellow and red stars.

To a great extent luminosity and colour are parallel attributes. Stars of a blue or white colour have the greatest absolute magnitudes, while the red are generally the least intense; yellowish stars, of which the Sun is an example, lie roughly half-way between these two. Colour goes with temperature; the higher the temperature, the more intense the radiation.

A closer study of the luminosities and the colours of the stars revealed to Ejnar Hertzsprung that not all stars of the same colour have similar luminosities, but that a few, especially among the red and yellow types, stand out on account of their relatively high luminosities. In two articles which appeared in 1905 and 1907 in *Zeitschrift für Wissenschaftliche Photographie* under the title 'Stellar Radiation', he came to the conclusion that, if the stars are to be classified according to their colours, it is necessary to differentiate between two sequences: in the one, successive steps from white through yellow to red signify a regular decrease in luminosity from 100 to 1/100, while, in the other, stars of all colours have luminosities of 100.

Hertzsprung expressed this in the following terms. The luminosity as an expression of the total radiation depends not only on the temperature of a star, but also on its diameter. The temperature merely determines the specific intensity of radiation emanating from one square centimetre of the surface of the star. This value must therefore be multiplied by the entire surface area measured in square centimetres to yield the total radiation. Hence the more luminous yellow and red stars must have very much greater diameters than the Sun; they are 'giants' in comparison with the 'dwarfs' which resemble the Sun.

Ten years before Hertzsprung the British astrophysicist

Norman Lockyer had suggested, on the basis of purely theoretical considerations, the existence of these two sequences. Within the framework of his cosmogony they represent either ascending or descending limbs of stellar evolution. A few years after Hertzsprung, in 1913, Henry Norris Russell correlated both the empirical discoveries and the cosmogonistic significance in the 'Giant and Dwarf Theory of Stellar Evolution'. This marks the actual birth of the *Hertzsprung–Russell Diagram* with its double significance, firstly as the correlation between colour and luminosity—in this form it is also known as the Colour–Magnitude Diagram—and secondly as a graphic representation of stars in various stages of evolution.

At that time the theory (founded on Lockyer) was that a star in its visible form began life as a red giant; then became hotter through contraction, and so with diminishing diameter and rising temperature passed through the stages of yellow and white giant until it reached maximum temperature and luminosity; from then on the star would begin to run down, as it were, since further contraction failed to yield enough energy to match that lost through radiation. In consequence temperature and luminosity would decrease and the star would become a dwarf.

The double nature of the Hertzsprung–Russell Diagram as a graph depicting both state and evolutionary stage has not always been clearly recognised, and misunderstandings have thus occasionally arisen over what represents a theoretical tendency and what represents the results of observation. Let us concentrate chiefly on the empirical side of the question, that is to say on the Colour–Magnitude Diagram, as a graphic representation of the various states, characterised through the two measurable values of colour index and luminosity, in which matter, in the form of stars, occurs in the cosmos.

In this two-dimensional diagram let us mark the sequence of colours (blue, white, yellow, red) from left to right along the horizontal axis; these correspond to the temperatures of the stars. Down the vertical axis we mark the luminosities in descending order. The Sun, a yellow star with an absolute magnitude of 5 (and a luminosity of 1), is marked with a dot a little to the right and below the centre of the diagram.

The discoveries made by Hertzsprung and Russell in the decade between 1905 and 1915 may be summarised as follows: From the evidence of luminosity, some of the stars are giants and others dwarfs. The luminosities of the giants are all about 100 and their colour makes little difference; they lie in a band more or less horizontally across the diagram from left to right. In the dwarfs luminosity and colour run parallel; in general we may equate the colour white (temperature 10,000°) with a luminosity value of 100; yellow (temperature 6,000°) with luminosity 1, and red (temperature 3,000°) with luminosity 1/100. Thus the dwarfs lie in a diagonal band from top left to bottom right of the graph.

Giants are apparently much less frequent than dwarfs; among the hundred or so stars in the immediate vicinity of the Sun there is, for instance, not a single giant. Similarly, the open clusters consist almost exclusively of dwarf stars, and only the occasional giant is to be found.

Slight differences in the spectra of giants and dwarfs were first noticed by Adams and Kohlschütter at the Mt. Wilson Observatory, and led to the laying down of criteria for distinguishing between the two types, as well as providing the groundwork for a two-dimensional classification of stellar spectra, by adding either the letter g (giant) or d (dwarf) before the Harvard classification. In this way the Sun is classified as dG2. These spectral criteria, whose physical origins we shall examine later, play a part in the photometric determination of distance as computed from apparent magnitude and luminosity.

The first schematic colour–magnitude diagram, with its two branches, giants and dwarfs, was soon extended. Among the stars in Hertzsprung's first list there is a white star whose luminosity is more than a thousand times smaller than that of a normal star, say Sirius or Vega; it is the companion of the double star O_2 Eridani. In 1914 Adams found the companion of Sirius to be another such white star of low luminosity. The high temperatures of these bodies as demonstrated by their spectra, and their low luminosities can best be explained if one assumes that their surface areas are very small; their diameters must be less than one-fiftieth that of the Sun. However, it also means that, in the case of Sirius' companion, a mass equivalent to that of our Sun is embodied in a globe of about the same size as the Earth. The density of matter in this 'white dwarf' must therefore be between ten thousand to a hundred thousand times greater than that of water.

In 1922 Eddington pointed out that such a concentration of matter—a thimbleful of matter from the companion of Sirius would weigh more than a hundredweight on Earth—was definitely possible in the light of what was known about the structure of the atom, since, bereft of their electron shells, the nuclei of atoms could be packed more densely. The physicist talks in terms of degenerate or collapsed matter to describe this state. The gravitational field at the surface of a star consisting of such collapsed matter is so strong that the spectral lines would show a relativistic red shift forty times greater than in the solar spectrum. In 1923, Adams was able to show that this was so in the case of the companion of Sirius.

The existence of a new type of star, the white dwarf, was thus confirmed, and could be entered on the diagram as a third type of state, quite apart from the giants and normal dwarfs. The dots indicating the white dwarfs lie along a band parallel to the diagonal band containing the normal dwarfs on the side of lower luminosity (1/100 to 1/10,000).

On the other hand, research by Shapley on the question of luminosity and colour in the globular clusters led to modifications in the field of giants. In contrast to the open clusters of the Milky Way system, in which the blue and white types of stars are invariably the brightest, and luminosity diminishes towards the red, the situation in the globular clusters is more or less the opposite. Here the red stars are brighter than the white, and there is a total absence of the blue type. On the colour–magnitude graph of the globular clusters, the dots representing the stars of this category form a band from the top right-hand side of the diagram (red stars of luminosity 1,000) running at a slight angle down towards the left as far as the yellow stars of luminosity 100, and at that point splits into two branches; one continues horizontally to the white stars, and the other curves downwards to merge with the dwarfs.

The characteristic difference shown in the correlation of colour with luminosity in the globular clusters, as compared with the open clusters, is one of the most telling arguments in favour of Shapley's scale of the Greater Galactic system. It contains the germ of the idea which Baade was to develop twenty years later, the differentiation between two types of stellar populations.

In the early twenties two questions were of paramount importance:

1. Is the region of the colour–magnitude diagram, which lies between the branch of the giants and that of the dwarfs, utterly devoid of stars, or are there some intermediate types, as for example red stars whose luminosity is equivalent to the Sun's?

2. How broad are the two branches, that is to say what dispersion is there in the actual degree of luminosity among giants and dwarfs of the same colour, taking into account any observational inaccuracies which are likely to exist in the observed distribution?

In about the year 1936 statistical investigations into the motions of yellow and red type stars pointed to the existence of a class of sub-giants which fill the gap in question. In due course, spectral properties were found which confirmed the theory, though it was still debatable whether these stars should rank as 'sub-giants' or 'super-dwarfs'. It was at about this time that people began to question the appellation 'dwarf' for the stars forming the diagonal branch, since there were no specific criteria for the white stars at the point where the horizontal branch of the giants cuts across the diagonal of the 'dwarfs', so that the stars of this kind could not be placed with any degree of certainty either in the one category or the other. Because of this dilemma a new term gradually came to be used, and so the stars which form this diagonal branch are now always known as 'main sequence' stars.

The second question, which is one of cosmic distribution, was solved through precise photometry of star clusters; a study of Præsepe with the astrograph at Göttingen (1935) yielded the information that all the stars which are definitely known to be members of this cluster lie on a single curve. Similar photometric studies of other clusters corroborated these findings, and showed in addition that the main sequences of individual clusters, when plotted on the colour–magnitude diagram, did not actually coincide, but instead lay side by side.

This threw a new light on a suggestion made by Trümpler (whom we have to thank for a vast general study of the galactic clusters) in 1927, namely that the clusters should be classified according to the form of their colour–magnitude diagrams. Trümpler found that the stars of a given cluster only ever covered a part of the general colour–magnitude diagram. In the case of the Pleiades, for instance, only the main sequence is occupied, but this stretches right up to the brightest blue stars, while in the graph of Præsepe the upper limit of the main sequence is with the white stars; there are no blue stars in this cluster, but on the other hand there are a few yellow giants. The situation in the Hyades is similar.

Whereas the first two decades had been devoted to the general consolidation of the Hertzsprung–Russell Diagram, the details of the diagram now became increasingly important. Considerations based on the theory concerning the internal constitution of the stars and of stellar atmospheres began to enter into the argument; astronomers began to sketch in lines on the diagram linking conditions of state which displayed some sort of common physical factors.

In 1925 Russell calculated that, according to Eddington's theory of the internal constitution of the stars, it would seem that the central temperature of all stars forming the main sequence was roughly of the same order, between 32 and 40 million degrees, and that the main sequence could thus be regarded as the isotherm of central temperatures. The significance of this in the field of cosmogony becomes fairly obvious when one remembers that the release of energy through the creation of more complex atoms from hydrogen requires extremely high temperatures, such as can be attained only in the interiors of stars.

In 1933 a young man named Bengt Strömgren, working on theoretical considerations concerning the chemical composition of stellar matter, in particular the hydrogen and helium content, arrived at the conclusion that, when stars with similar hydrogen content are plotted on the colour–magnitude diagram, they form lines which follow the main sequence from the bottom right towards the top left and then turn to the right in the upper portion towards the giants. The lines of equal hydrogen content form a series of parallel curves; a lower hydrogen content corresponds to a displacement of the curve towards the right-hand side of the diagram, in a direction perpendicular to the main sequence.

The resemblance of these curves to the colour–magnitude graphs of the open clusters implies that in these clusters one is dealing essentially with groups of stars having similar chemical compositions, and that Trümpler's classifications, in which main sequences also show a sideways displacement on the colour–magnitude diagram, are indeed groups with different hydrogen contents. The assumption that evolution in stars goes with diminishing hydrogen content seems appropriate; furthermore it is a clue regarding the ages of the star clusters.

Briefly, therefore, by 1940, we had the following picture of the Hertzsprung-Russell Diagram in its primary function as a general graph showing the conditions of state and the various forms in which matter exists in the Milky Way system as stars (figure 10).

There is the main sequence of stars which occupies a diagonal branch running from the top left of the diagram (B-type stars with luminosities of 10,000) down to the bottom right (M-type, luminosity 1/1,000); the width of this branch depends on chemical composition. The Sun is a star of the main sequence (G-type, luminosity 1). Above the main sequence there are the super-giants, giants and sub-giants corresponding to luminosities ranging from 1,000 to 100,000, from 100 to 1,000, and from 5 to 75 respectively. The super-giants cover the whole range of colours from blue to red, while the giants and sub-giants extend only from yellow to red. Between the main sequence stars and the zone of giants and sub-giants there is a relatively empty area which is known as the Hertzsprung Gap. A- and F-type stars, with luminosities between 10 and 100, and temperatures between 10,000° and 6,000°, do not seem able to exist as stable structures. It can hardly be a coincidence that variable stars of high luminosity (especially the Cepheids) happen to come within this range of temperature and luminosity.

Below the main sequence there are three distinct areas. The yellow and red sub-dwarfs occupy a small strip parallel to the main sequence; luminosities of these stars are about a third of those for main sequence stars of corresponding colour. The white dwarfs also lie along a strip parallel to the main sequence, but their luminosities are a hundred times smaller. At magnitude 1 are to be found extremely hot blue stars, such as occur at the centres of planetary nebulæ, and novæ in their quiet states before or after an outburst.

Although there appears to be great diversity in the universe as a whole, the colour–magnitude diagram shows that this is confined to fairly specific areas on the graph, particularly in differences shown by the graphs of the open galactic clusters and the globular clusters. The roughly qualitative distinction between two types of star populations, as introduced by Baade, takes this into account.

Finer distinctions between the colour–magnitude diagrams within groups of similar populations—open clusters, Population I, and globular clusters, Population II—underline the fact that these two parameters, luminosity and colour, are not sufficient in themselves to allow a full assessment of the physical state of a star. What is needed is a third parameter, which gives a numerical value for some aspect of chemical composition, as for instance the hydrogen content. It became generally accepted that the chemical composition of the stars varied.

It was at this time that Bethe and von Weizsäcker put forward the theory concerning the generation of atomic energy in the interiors of stars. This made it possible to explain the varying chemical compositions of the stars as a result of elemental changes brought about by nuclear processes—fusion of hydrogen atoms to form helium and other more complex atoms—in the course of their evolution. In addition, differences displayed by the graphs of individual groups could be used to determine the ages of the groups concerned.

In the history of the Hertzsprung–Russell Diagram during the last twenty years, this line of thought emerges more and more. The efforts of astronomers are directed towards tracing lines on the colour–magnitude diagram linking stars of the same age. The main sequence is split into a succession of rows denoting particular chemical compositions. Attempts have been made to try to deduce an 'initial sequence' of stars containing the original matter. It follows that the degree of displacement of the main sequence of a given group of stars relative to this 'initial sequence' can be used as a measure of age.

These manœuvres with the concept of age are not without certain dangers, insofar as one tends to forget that the age of a star, or group of stars, or system of stars is not something which can be observed directly in the same way as magnitude, or temperature; it can never be more than a theoretical consideration, or at most an apparent explanation of some observed phenomena.

If the two parameters, colour and luminosity, are inadequate for making empirical distinctions between various stellar states, and a third parameter has to be brought into the calculation, then the criterion of age is not the one to choose; far better to look to the spectrum to provide a numerical value which expresses the chemical composition. This is exactly what was proposed by Daniel Chalonge of the Institut d'Astrophysique, who demonstrated a relief model illustrating a three-dimensional concept of state in a symposium on the Hertzsprung–Russell Diagram at the Astronomical Congress in Moscow in 1958. His three-dimensional model replaces the two-dimensional colour–

magnitude diagram. Hertzsprung himself, then 85 years old and still playing an active part in astronomy, was present at the demonstration, which became a triumph for the father of the Hertzsprung–Russell Diagram.

7 · Stellar Atmospheres

In the minutes of the *Gesellschaft der Wissenschaften* of Göttingen, there appeared in 1906 an article by Karl Schwarzschild on 'The Equilibrium of the Solar Atmosphere'. The article covered only a few pages, but it introduced the idea of balanced radiation, as well as the first physical explanation of the structure of the Sun's outer layers. It seemed to provide the answer to two of the most important phenomena which had been observed: the decrease in the apparent brightness of the solar disc towards the limb, and the sharp outline of the disc itself. From this article grew theories concerning the atmospheres of the stars. The mathematics of this were explained in a comprehensive paper by Schwarzschild which is recorded in the minutes of the Prussian Academy of Sciences (1914) in Berlin and entitled *On Diffusion and Absorption in the Solar Atmosphere*.

1938 saw the publication of a book by Albrecht Unsöld, *The Physics of Stellar Atmospheres*; the second edition of this (1955) contained 800 pages, and a bibliography which included as many as 2,000 titles, which shows just how much this field of research has grown.

The whole idea is based on spectra and what they can tell us about stellar radiations. The spectrum of the Sun consists of a continuous background, which the human eye sees as that familiar band of colours, from red through orange, yellow, green, blue, and indigo to violet. Not so apparent are the numerous dark lines which overlie this background. These are named after Fraunhofer, who used them as demarcations of wavelengths. The most prominent of these lines he identified with letters of the alphabet.

According to the explanations given by Bunsen and Kirchhoff, these dark lines on the continuous spectrum occur when the radiations emitted from a glowing solid pass through a layer of gas; in fact, they occur in exactly those places where the spectrum of the gas would show bright emission lines. This reversal of the spectral lines, as it is called, forms the basis of chemical spectral analysis, and was first used by Bunsen and Kirchhoff in 1859 in a study of the Sun.

The explanation of the Fraunhofer lines in the solar spectrum as absorption lines led to the idea that the Sun is an incandescent globe—whether this is solid, liquid or gaseous is immaterial at the moment—which emits a continuous spectrum, and that this globe is surrounded by a gaseous mantle, which gives rise to the absorption lines.

During a total eclipse of the Sun, the spectrum changes abruptly as soon as the Moon covers the entire luminous disc of the Sun; the continuous background fades, and, in the places where the dark Fraunhofer lines usually come, there appear bright emission lines. The 'flash spectrum', as it is called on account of its rapid appearance, is, in fact, the spectrum of the gaseous surround.

It is possible to distinguish three layers in the Sun's atmosphere: the 'photosphere' which emits the continuous spectrum, the 'reversing layer' which gives rise to the absorption lines, and the 'chromosphere' which is responsible for the flash spectrum.

In the simplified model of the solar atmosphere which Schwarzschild described in 1906, he suggested that the radiations emanating from the interior enter the bottom of the atmosphere with an intensity which is distributed equally in all directions; is absorbed in its passage through the solar atmosphere and re-emitted, until it finally emerges from the surface —only now, its intensity is different in different directions. The greater the inclination of the ray from the vertical, the more atmosphere it has to pass through, and the more it is weakened in the process.

If the atmosphere is to remain mechanically stable, then, at any given level, the weight of matter has to be borne by the pressure of the gas below. Because of this there is uniform decrease in density with altitude, depending on gravity and the molecular weight of the gas.

Schwarzschild next brought a new theme into the considerations, namely that the Sun's atmosphere must also be in a state of equilibrium so far as radiation is concerned; this would mean that the stream of radiations passing through the atmosphere from below suffers no change in strength. Since this flow of radiations depends on temperature, this must decrease with height corresponding to the density, if the radiative balance is to be maintained.

The radiation stream which passes through is characterised by the 'effective' temperature; according to the Stefan–Boltzmann Law the intensity of radiation is proportional to the fourth power of the effective temperature. From the solar constant, which is a measure of the energy stream reaching the Earth, roughly 2 cal./cm^2 per minute, we can calculate an effective solar temperature of nearly 6,000° Kelvin.

Schwarzschild's simple model tells us two things about the Sun: in the first place it tells us that temperature and density decrease from the interior outwards, and secondly it yields a formula for the darkening of the limb, that is to say the diminishing brilliance from the centre of the disc towards the edge. An effective temperature in the order of 6,000°K would entail a surface temperature of 5,050°K; and the intensity at the limb is about two-fifths that at the centre.

In this model, however, the upper limit of the solar atmosphere lies, theoretically, at infinity, but the decrease in temperature and density is so rapid, that in reality the edge of the Sun appears sharply defined. If the altitude is calculated from the level at which the temperature is 6,000°K, it is necessary to rise only 400 km. above this level to reach a zone where the temperature is 5,051°, and where the density is only one-thousandth that

85

at the starting level. A depth of 400 km., however, represents no more than one two-thousandth of the Sun's radius, i.e. half a second of arc. Consequently, the edge of the apparent disc of the Sun appears sharp within a tolerance of less than half a second of arc.

Schwarzschild's explanation is merely a first approximation. In order to get at the real facts, we must put aside the simplified assumptions. The first task is to solve a purely mathematical problem which is known as radiative transfer. The decisive steps in this direction were taken in the decade beginning 1920. In 1929, E. A. Milne gave the first comprehensive description in the *Philosophical Transactions of the Royal Society*: 'The Structure and Opacity of a Stellar Atmosphere'. Then in 1934 there appeared a book by Eberhard Hopf: *Mathematical Problems of Radiative Transfer*.

Apart from the mathematical problem of radiative transfer, there is also the physical problem of the absorption coefficient, a decisive factor in radiative transfer. Atomic theory supplies the numerical values for the atomic absorption coefficients of the individual elements, and from this one can calculate the mean absorption coefficient of matter found in a stellar atmosphere. In about 1925 the first data became available from which model atmospheres of different compositions could be computed.

The structure and radiation properties of such a theoretical atmosphere in a state of radiative equilibrium are determined through

1. The effective temperature, as a measure of energy flux;
2. gravitation, which governs the mechanical equilibrium;
3. the relative frequency of the various kinds of atoms, from which the mean molecular weight and the progression of the absorption coefficient with wavelength can be determined.

As observational proof of the theory, both the limb darkening and the distribution of energy in the continuous spectrum may be cited.

Measurements of the darkening at the limb will be of

86

use in respect of a decision in favour of one or other of the proposed models only if the falling off in brilliance can be measured accurately so far as the last $\frac{1}{2}\%$ of the solar radius is concerned, that is to say within a rim no more than a few seconds of arc in width. Again and again attempts were made to take sample readings within small areas of the Sun's disc along a chosen diameter, only to fail, because stray light and atmospheric disturbances blurred the outline of the apparent solar disc. During solar eclipses conditions are far more favourable, since at such times the Moon moves across the face of the Sun like a shutter; it lies outside the atmosphere and is therefore able to mask the various layers one after another. Hence it was not until eclipse observations of the last ten years or so that results could be obtained which were sufficiently accurate to be of use in deciding between the various model structures of the solar atmosphere, which had been suggested at various times.

From the energy distribution observed in the continuous spectrum of the Sun, it is possible to deduce the effective temperature, provided surface gravity and the relative frequency of the elements are known. In order to understand clearly what is meant by the adjective 'effective', we need to consider the various concepts of temperature which are used to characterise radiations.

Let us begin with Planck's radiation law for the ideal 'black body'. Such a black body is one which completely absorbs all radiations which fall on it. It can be imagined as a hollow space enclosed on all sides by walls absorbent on the inside. Any radiation which enters through a small hole in the wall will be completely swallowed, because, even though the absorbing powers of the interior walls may not be absolutely perfect, any residual radiation which is reflected will strike another wall where more of it will be absorbed. Thus total absorption is effected, and none of the radiation ever re-emerges. In this sense the term 'black' is not necessarily synonymous with 'dark', but means 'non-reflective'. It is rather as if the whole of the hollow space is filled with radiations whose intensity and spectral distribution depend solely on the absolute temperature. The laws of radiation formulated by Planck in 1900 show the relation between the intensity and the temperature of the black body, so that, if the second of these two factors is known, the first can be worked out. Temperatures calculated on this basis are called 'radiation temperatures'.

Planck's Law implies that in an ideal black body the radiation temperature can be calculated solely from the ratio of intensities of two different wavelengths; there is no need to measure the absolute intensity, nor does it matter which two wavelengths are chosen.

Astronomers now had at their disposal an equation for determining stellar temperatures. If the stars behave like black bodies, then their surface temperatures can be calculated from their colour indices; the colour index represents the intensity ratio between two wavelengths. Astronomers now began to tackle the problem from this aspect, and in this sense they referred simply to *the* temperature of the Sun or a star, without qualifying the term.

It soon became apparent that in reality conditions were not quite so simple. If one bases the calculation of the solar temperature on the total radiation, then with a solar constant of 2 cal., one arrives at an effective temperature of 5,800°C. On the other hand, if the calculation is based on the relative energy distribution in the spectrum, the colour temperature of the visible portion of the spectrum reaches a value of more than 7,000°, while in the short-wave region the temperature is less than 5,000°. Finally there is still the difference between the centre of the disc and the edge; the colour temperature taken as a mean of the whole disc is several hundred degrees lower than at the centre. Which of these values should therefore be regarded as *the* temperature of the Sun?

In the case of the stars, a similar situation arose in the contrast between two series of spectro-photometric observations. One of these—published in 1909 by Wilsing and Scheiner of the Astrophysical Observatory in Potsdam—consisted of the visual part of the spectrum, while the other, by Hans Rosenberg in Göttingen (1914), concentrated on the photographic portion.

Rosenberg's results for the white stars were very much higher, and for the red stars considerably lower than those of Wilsing and Scheiner. The temperature scale of the stars calculated from the photographic spectrum has a wider range than the visual; in fact the ratio is 16 : 10. Which scale, therefore, is the correct one?

An observation made by Johannes Hartmann at the Potsdam Observatory in the course of some photographic photometric studies of the spectra of hot stars pointed the way to a solution of this apparent contradiction; he described what he had observed as a depression in the ultraviolet range. He found that the intensity distribution in stellar spectra did not match those of a black body; Planck's Law cannot be applied to the stars. Despite the fact that in the interiors of the Sun and the stars conditions may obtain which will permit genuine black radiation of a certain temperature to be generated, these radiations are modified in their passage through the outer layers; hence the energy curve becomes distorted on emergence from the photosphere. The task of an exact theory of stellar atmospheres is to determine the amount of distortion.

This then was the situation in about 1915, when the theory of stellar atmospheres was still in its infancy. The urgency of finding a solution to this problem becomes clear when one considers that the temperatures are of interest not just for themselves as a measure of a particular physical state, but they also play a part in determining other stellar characteristics.

In accordance with the laws of radiation, it is possible to use effective temperature for calculating the energy which is emitted from each square centimetre of the surface in the form of radiation. If the total amount of energy radiation is known (in the case of a star, this is equivalent to the true luminosity), and the temperature (that is to say, the surface brilliance), it is possible to calculate the area of the radiating surface, and from this the radius of the star can be found. If then the mass of the star is also known we can calculate the mean density of the matter from which the star is composed. Next to temperature, density is probably the most important physical factor, for it tells us something about the material of which the star is composed, namely whether it is likely to be gaseous, liquid or solid.

From the stellar temperatures which Wilsing and Scheiner determined in 1919, they calculated the radii of the stars concerned, among them some of giants. A young astronomer at the Berlin Observatory named Bernewitz used the same temperatures to calculate the densities of binaries and, in this way, stumbled on a high density for the companion to Sirius in 1921. When he tabulated his results, he marked the entry with two bold question marks, and added the note that one could arrive at a 'sensible' result only if one ascribed to this star a very much lower temperature than its observed spectrum (Harvard Classification A) would suggest. If the temperature were 2,000° instead of 10,000°, the density value would sink from 88,000, as calculated, to 5.

This state of affairs at the beginning of the twenties posed a twofold question for observation and theory:

1. Exact measurement of the distribution of absolute intensity in the continuous spectrum of the Sun and of stars of all spectral types over the broadest possible range of wavelengths.

2. Calculating the theoretical spectra of model atmospheres having various effective temperatures and compositions.

We have already dealt with the most important point with regard to the second task, the need to develop a comprehensive theory of stellar atmospheres going beyond Schwarzschild's approximation. On the specific problem of the theoretical spectrum there are a few points relating to historical background that I would like to add at this stage.

The relative spectral intensity distribution, that is to say, the shape of the energy curve over the wavelengths, is determined to a considerable extent by the range of the continuous absorption coefficient; the latter can be calculated from the atomic absorption coefficients according to the relative abundance of the various kinds of atoms which constitute the atmosphere.

The characteristics of hydrogen dominate the spectra of hot

stars. Moreover, the hydrogen atom, which consists of a nucleus and only one electron, is a comparatively simple subject for the theory of atomic absorption coefficients. Hence it is hardly surprising that the first theoretical spectra of stellar atmospheres should be calculated on the basis of pure hydrogen. This was done by a young British physicist named McCrea, who studied at Göttingen at the beginning of the thirties. These calculations began to shed some light on the differences between the temperature scales of Wilsing on the one hand and Rosenberg on the other.

A little later, Anton Pannekoek in Amsterdam and Albrecht Unsöld in Kiel published their results of calculations, in which the admixture of atoms other than hydrogen—collectively described as 'metals'—was taken into account. These calculations varied according to the percentage of heavy atoms assumed to be present. Whereas Unsöld assumed the ratio of hydrogen to metals to be 50 : 1, Pannekoek suggested that there was a considerable preponderance of hydrogen in comparison to heavier atoms in the ratio of 1,000 : 1. Which particular relation between colour temperatures (deduced from various sections of the continuous spectrum) and effective temperatures gives a more reliable picture of conditions could not be decided with the spectro-photometric evidence then available.

The question of the relative frequency of hydrogen is one to which we shall return in connection with the internal constitution of the stars. At this stage, suffice it to say that from about 1935 on there has been increasing evidence in favour of Pannekoek's higher value for the ratio of hydrogen to metals.

Rupert Wildt's ideas on the mechanism of absorption plays an important part; these are founded on the discovery of the negative hydrogen ion (1939), and the positive molecular ion of hydrogen (1947). Hence it is possible to calculate that portion of the absorption coefficient which Unsöld attributed to the metals.

In 1940 Bengt Strömgren postulated his first theoretical model of the solar atmosphere, basing his calculations on the proportions in which the various atomic types were present. His assumptions concerning the composition of the absorption co-efficient were obviously fairly accurate, and provided a satisfactory description of the continuous spectrum. Since that time a number of further calculations have been carried out with slight variations in the basic assumptions and by different methods; as a result a growing number of atmospheric models of all spectral types is now available.

Spectro-photometric observations were developed along with the development of the theory, to the benefit of both. From the series of observations, which were taken in hand in various places during the twenties, to try to determine the temperatures of the stars from the continuous spectrum, there were two especially significant contributions for testing the theory and for providing a means of deducing an absolute stellar temperature scale. In the 'Göttingen Temperature Programme' a fundamental system of thirty-six stars was achieved, and since 1940 this has been used extensively to integrate other stars into the scale of stellar temperatures. An improved system based on more recent photo-electric data should shortly appear.

In Paris, a programme was undertaken by Barbier and Chalonge, which was at first concerned primarily with the investigation of the ultraviolet part of stellar spectra. Since the start of this programme its scope has been considerably widened by extending research into the realm of the longer wavelengths, and by the inclusion of stars of every conceivable type. From these investigations there emerged the three-dimensional classification of stellar spectra proposed by Chalonge, which allows for the fact that the state of a stellar atmosphere can be theoretically determined by means of three parameters, namely temperature, gravity, and the relative frequency of component atoms.

By comparing the results of observation with theory a somewhat unsatisfactory situation arose in about 1955. In the continuous spectra of B and A type stars deviations were noticed which the existing theory was unable to explain. However, thanks to greater accuracy of the photo-electric observing methods, as compared with the photographic-photometric methods, a solution to the problem is in sight. The apparent anomalies probably result from a combination of systematic errors, which occur partly through faulty assessment of the photographic plates, and partly through the blurring of absorption lines consequent upon insufficient spectral resolution.

This brings us to a point which plays a decisive rôle when observed and theoretical spectra are compared. It hinges on the question of how genuine and accurate a continuum can be achieved through observation, and whether the actual observational spectrum is generally identical with the ideal spectrum, whose intensity distribution is based on theory.

So far, we have ignored the absorption lines which form a part of an actual spectrum. They affect our considerations in two ways. In observation they make it difficult to record an undisturbed continuum between the lines if there is insufficient spectral resolution; in the theory, they necessitate the introduction of corrections which will allow for changes through absorption in the radiation flux. Both these factors, the correction through lack of spectral resolution, and revision of the theory, presuppose a knowledge of the intensities of these lines and their distribution through the spectrum. In order to achieve a comprehensive theory of stellar atmospheres, it has become necessary to supplement the theory of the continuous spectrum by means of a theory concerning absorption lines.

To deal with this aspect would, however, require a complete chapter on its own account, and goes well outside the scope of what I have tried to outline here. I shall have to content myself with the remark that the theory of stellar atmospheres, so far as it concerns the interpretation of continuous spectra and absorption lines, is nevertheless sufficiently comprehensive to provide significant data, in particular the effective temperature, which will describe the state of and the processes occurring in stellar interiors.

8 · The Internal Constitution of the Stars

In the volume entitled *Physics of the Cosmos* of the textbook on physics by Müller-Pouillet (1928), there is a chapter called 'The Stars as Radiating Gaseous Bodies'. At the time, I wrote in the introduction: 'That we can nowadays talk in terms of the thermodynamics of the stars is due to the development which has taken place in astronomy during the last two decades under the influence of an increasing number of physical considerations. The way had, of course, already been paved, and it would be unfair not to remember the valuable contributions made to theoretical astrophysics by Helmholtz, J. H. Lane, A. Ritter, W. Thomson, and others. Meanwhile, one would not be far wrong if one were to regard as the beginning of this new development Schwarzschild's short discourse (1906) *On the Equilibrium of the Solar Atmosphere*, in which the concept of radiative equilibrium was first mentioned in astronomical literature. The book *Gaskugeln* by R. Emden, which came out at much the same time, may be regarded as dealing exclusively with convective equilibrium, but this in no way detracts from its importance, since more recently it has formed the basis for new lines of investigation.'

'Every star is a radiating sphere of gas'—this apparently simple answer to the question of what is meant by a star, hides a host of problems.

Schwarzschild's hypothesis applies only to the outer shell, the Sun's atmosphere. Eddington went further: he considered the star as a whole. His first work *On the Radiative Equilibrium of the Stars* was published in 1916, and his second *The Internal Constitution of the Stars* in 1926. The following words from the foreword epitomise the situation. 'The book was written between May 1924 and November 1926. Time was occupied by a number of minor investigations, made to fill the gaps that disclosed themselves as material was brought together. Anyone writing on a theme which many workers are actively investigating is liable to find his pen unable to overtake the rate of growth of the subject. During the above-mentioned period the theoretical papers on stellar constitution in the *Monthly Notices* alone amounted to more than 400 pages. It has been still more difficult to cope with modifications and progress in the theory of the atom, on which astronomical developments must rest. As we go to press a "new quantum theory" is arising which may have important reactions on the stellar problem when it is fully developed.'

The stars are spheres of gas in a state of radiative balance. What does this statement imply? In the interior of a star, mechanical equilibrium exists so long as the material particles are immobile, neither sinking towards the centre, nor rising towards the surface. Thermal equilibrium exists, so long as no region of the interior undergoes a unilateral change of temperature, and radiative equilibrium exists if the energy absorbed by a given volume of matter is equal to that which it radiates.

If we wish to apply these concepts of equilibrium to the problem of the interiors of stars, then we must consider what are the forces which determine mechanical equilibrium, and by what means energy is transferred from one place to another. Heat, we know, can be transferred by conduction, convection and radiation.

When one end of a metal rod is heated, the heat energy travels along the length of the rod; if the heat source is working continuously, the heat in the rod will begin to balance itself out with a certain temperature gradient from one end to the other.

When a heater warms the air in a room, heat is conducted to a thin layer of air which is in immediate contact with the heating

95

surface. The heated air expands and rises, and so a convection current sets in. In this way warmed masses of air are able to reach other parts of the room, and there is a continual mixing of warm and cool air masses. Here there is also a state of balance with a steady decrease in temperature in all directions away from the heater. For obvious reasons, this kind of equilibrium is called convective. The gaseous spheres of the classical theory, as they are described in Emden's book, are in a state of convective equilibrium.

However, the heater also radiates heat, and in this way transfers heat energy to more distant places without necessarily warming the intervening air masses to any appreciable extent. In a room heated by means of ceiling radiators a sort of radiative equilibrium can occur, with a vertical temperature gradient of stable horizontal layers of air.

The stars are 'gaseous spheres'. One can demur at the idea of describing the state of matter in the interior of the Sun as a gas, if one remembers that the mean density of solar material is rather higher than that of water. But since the work of Hertzsprung and Russell, we know that some of the stars are giants in which the density of matter is very much less than that of the air of our atmosphere. So far as these stars are concerned, therefore, the gas-sphere idea can be said to apply.

Even Eddington considered and intended his theory to apply solely to the giants; for the Sun and all stars of the main sequence, with mean densities ranging between 1/10 and 10, an equation of state other than that of an ideal gas would apply. It was only from about 1923 onwards that it became at all clear that matter in the interiors of the stars, despite the high density, does, in fact, behave like an ideal gas, and also why this is so.

The characteristics of the gaseous state include the fact that the distances between the atoms are considerably greater than the diameters of their electron shells. Consequently the atoms are free to move; the resistance which a given quantity of the gas offers to any reduction in volume (i.e. an increase in pressure) depends on the energy with which the atoms in their movement impinge on one another and on the walls of the container.

The kinetic theory provides the link between pressure, temperature and density in the form of an 'equation of state', so that, if two of the factors are known, the third can be calculated. A material constant comes into this equation of state, and this is characteristic for a given gas; it is called the 'molecular weight', and is the mass of the smallest natural particle of that gas (molecule, atom, or even only a part of an atom) expressed in terms of the mass of a hydrogen atom. The structure of gaseous globes varies with molecular weight.

What, then, is the state of matter in the interiors of the stars, where we may anticipate temperatures of many millions of degrees? At such temperatures the atoms collide so violently, that they become more or less completely denuded of their electron shells; the physicist refers to this process as 'ionisation'. The number of electrons which surround the atomic nucleus increases as the 'atomic weight' of the substance. Oxygen, whose atomic weight is 16, has 8 electrons; calcium, whose atomic weight is 40, has 20; iron, with an atomic weight of 56, has 26 electrons; uranium, atomic weight 238, has 92 electrons. In the lighter elements the atomic weight is equivalent to twice the number of electrons, whereas in the heavier elements the value is slightly more than double the number of electrons.

When the bond, which holds the electrons and the nucleus of an atom together, is broken, and these components are free to move independently, the gas changes its character. The 'molecules', if we may retain this term for the smallest particles—in this case, bare nuclei and electrons—are millions of times smaller in diameter than when the electron shells are intact. Consequently they are able to pack more densely before they lose their mobility through contact with other particles. In a given volume of such quasi-molecules, electrons far outnumber the occasional stripped nuclei, though almost the entire mass is concentrated in the latter.

Let us try to imagine a Sun which is composed entirely of oxygen, which has an atomic weight of 16; each atom consists of eight electrons and one nucleus, making nine components in all, among whom the mass of the oxygen atom (16) is distributed.

This means that the mean molecular weight of the gas is 16/9, i.e. 1·8. The corresponding value for iron is 2·1 (mass 56, number of components 27), and if the Sun were composed entirely of uranium, its molecular weight would be 2·6 (mass 238, number of components 93). Only for hydrogen and helium is the mean molecular weight markedly below a value of 2 under conditions of complete ionisation: hydrogen 0·5 (mass 1, number of components 2), helium 1·3 (mass 4, number of components 3).

Thus matter in the interiors of stars exists in a very simple form. The wide range of atomic weights of non-ionised elements, from hydrogen = 1 to uranium = 238, contrasts with the narrow margin for the electron gas, 0·5 to 2·6, and this is reduced still further if neither hydrogen nor helium is counted, 1·7 to 2·6. Although the presence of the various elements in the atmospheres of the Sun and other stars can be detected by means of their individual spectral properties, these tend to be largely destroyed under conditions which obtain inside the stars, where atoms are stripped of their electrons. Basically we are concerned with only two kinds of particles, electrons and atomic nuclei, which mix to form an 'electron gas' whose mean molecular weight is approximately 2, so long as neither hydrogen nor helium is present in any large quantity.

This idea was first expounded in detail in 1924 by John Eggert, a photo-chemist, in a short article in the periodical *Die Naturwissenschaften*. Until that time, Eddington had been working on the assumption of a molecular weight of 56, the atomic weight of iron, and in consequence had run into difficulties when comparing theory with practice. These difficulties were overcome once the value for the molecular weight in the equation of state was reduced from 56 to 2·2, as Eggert had suggested.

The electron gas inside a star is subject to intense irradiation; this has the characteristics of genuine black radiation, so that Planck's Law of radiation can be applied. At temperatures in the order of 10 million degrees, which are likely to be encountered, the maximum energy curve lies at 3 Å, that is to say in the region of the soft X-rays. According to the concepts of quantum theory (1923), the nature of radiation can be explained as waves, as well as corpuscular; it is thus permissible to regard the radiation field in the interior of a star as a gas, whose component molecules are energetic X-ray quanta. This 'photon gas' behaves exactly like an ideal gas, having a molecular weight, which may be calculated from its radiation energy by means of the mass-energy equation. Just as in the case of a normal gas, therefore, it is possible, by means of temperature and density, to calculate the pressure of the photon gas— the light pressure or radiation pressure.

If hydrodynamic equilibrium is to exist in the interiors of stars, then the forces which act on a particle of matter must maintain a stable balance. These forces are gravity, which tends to draw the particle towards the centre, in other words the weight of the matter, and, opposed to this, the resistance of the gas to compression, i.e. the pressure which results from two factors, the gas pressure of the electron gas, and the radiation pressure of the photon gas. The ratio, in which gas pressure and radiation pressure contribute towards the support of the weight of the material particles, is represented by a new factor which Eddington introduced into the theory of the internal constitution of the stars. It depends on the mass of the star and on the molecular weight of the matter of which the star is composed.

So far as the equilibrium of heat in the star's interior is concerned, conduction and convection play no appreciable rôles; the transfer of heat from the interior to the exterior is entirely effected through radiation. Hence, the structure of a star in a state of radiative equilibrium depends principally on the absorption coefficient of the stellar matter, since this governs the radiative transfer. In contrast with the atmosphere described by Schwarzschild, which is subject only to a constant radiation flow emanating from the interior, energy is also being released in the star's interior. The location of the energy sources, that is to say, whether they are evenly distributed, or whether they are more concentrated at the centre, plays a part in determining the structure.

In 1926, in a lecture on *Stars and Atoms*, Eddington, always a

master of lucid and readily understandable explanations, described conditions inside the stars in the following terms:
'Imagine a confused jumble of atoms, electrons and ether-
waves—severely damaged atoms rushing around at velocities
exceeding 100 km./sec., their normal complement of electrons
having been torn from them in the hurly-burly. The lost
electrons travel a hundred times faster, trying to find new resting
places. Let us try to follow the path of one of these. There is
almost a collision when an electron approaches an atomic
nucleus, but then with increased velocity it hurtles past in a
sharply curved path. Occasionally the electron takes a sideways
leap along its path, and then goes rushing on with increased or
diminished energy. After having lived through about a thousand
narrow escapes, all within a matter of a thousand-millionth of a
second, the headlong rush ceases with a more violent sideways
leap than usual. The electron has been securely captured by an
atom, but, no sooner has it taken its place, than an X-ray
strikes the atom. The electron absorbs the energy of this ray,
and immediately flies off to new adventures.

'And what emerges from this hurly-burly? Very little. For all
their haste, the atoms and electrons do not really achieve anything; they merely change places with each other. The ether-
waves are the only members of this society who manage to
create anything of lasting value. Although they appear to be
flying about indiscriminately in all directions, they do, on the
whole, gradually work their way outwards. The atoms and
electrons, however, are prevented by the force of gravity from
reaching the exterior. Slowly the pent-up ether-waves percolate
outwards, as through a sieve. The ether-wave travels from one
atom to another, sometimes forwards, sometimes backwards,
then hurls off in a new direction, and in the course of all this it
loses its identity, but survives in its successor. With luck it will
not be all that long (say, between 10,000 and 10 million years or
so, depending on the mass of the star) before it succeeds in
reaching the surface layers. As a result of the lower temperature
of this region, it now changes its character from X-ray to light
ray, though this is a gradual process involving a slight change

with every rebirth. At last it comes so near to the surface that it can make its escape, and wander through space for several hundred years or so without being disturbed. Perhaps, at long last, it will arrive at a distant world where an astronomer just happens to be lying in wait for it and having captured it with his telescope tries to wrest from it the secrets of its origin.'

For his first calculations Eddington lacked quantitative data both for the absorption coefficient as well as for the generation of energy. Like Schwarzschild in the case of the Sun's atmosphere, he had to rely on some simple basic assumptions. In this way Eddington's standard model of a star came into being; it was difficult to assess how accurately his ideas represented the actual conditions inside a star with regard to the ranges of pressure, temperature and density at various depths. Even in this early phase of the theory, one general tenet emerged which is now known as the *Russell-Vogt Theorem*.

This theorem states that the internal constitution of a star is uniquely determined through its mass and chemical composition. It is expressed in the Mass–Luminosity Relation proposed by Eddington, which establishes the relation between the two empirical values, mass and luminosity, which can be obtained through observational means; in this way it provides a means of putting theory to the test, provided that the mean molecular weight (i.e. the chemical composition) and the opacity (i.e. the absorption coefficient) are known and included in the theoretical considerations.

In the decade from 1923, during which time the new quantum theory developed, and gradually furnished astronomers with the necessary experimental and theoretical data, the following were probably the main points of consideration:

1. The kind of stellar models which were theoretically permissible within the framework of certain limiting conditions, and differing assumptions as to the nature and distribution of the energy sources, including the effects of a possible rotation. The 'point-source model' gained predominance insofar as the realisation prevailed that the only feasible energy source in the

interiors of stars which could be considered would be atomic, and that these sources could become effective only in the innermost cores of the stars under conditions of extremely high temperatures.

2. The relative frequency of the elements which compose stellar matter, and in particular the amount of hydrogen present, which has an important bearing on the molecular weight of the electron gas. The theoretical mass–luminosity relation can be satisfied by two divers assumptions for the hydrogen content; for example, the mass of the Sun could consist either of one third hydrogen, or alternatively 99% hydrogen.

The second of these assumptions seemed so improbable to the astronomers of the day that, even as late as 1932, Bengt Strömgren had this to say on the subject of the conclusions to be drawn from the mass–luminosity relation: 'The characteristic distribution of stars in the Hertzsprung–Russell Diagram occurs because 1. The majority of stars have a hydrogen content equivalent to one third of their masses, and therefore lie along the line of the main sequence and its extension into the branch of the giants, and that 2. Large mass seems to be possible only where there is a high hydrogen content.' The possibility that the Sun and also all the normal main sequence stars could consist predominantly of hydrogen was seemingly not worth considering. The realisation that almost all the matter in the universe exists in the form of the lightest atoms, hydrogen and helium, and that the heavier atoms constitute no more than an infinitesimally small proportion, began to dawn only in the next decade.

In this first stage in the development of the theory of the internal constitution of the stars, which came into being through the introduction of radiative equilibrium and radiation pressure, the number of stellar models which fitted observations became more limited, insofar as it was assumed that the composition of the electron gas in the entire interior of the star was homogeneous. A decision between the various models offered by theory was not possible on observational grounds so long as no linking factors regarding the composition of stellar matter were available through independent physical data and astronomical observation.

The dilemma was similar to that encountered at about the same time in connection with the continuous spectra of the stars. In this the relation between colour temperature (which could be deduced from observations) and the effective temperature (the decisive factor regarding the state of matter in stellar interiors) could also be determined by means of the chemical composition of the atmosphere, especially the relative frequency of hydrogen compared with the heavier elements. The choice lay between Unsöld's assumption (50 : 1), and Pannekoek's (1000 : 1).

In the years between 1930 and 1940 the ideas were born which were eventually to produce a concept of the internal structure of the stars which was to a great extent free from pure speculation. These ideas included:

1. A comprehensive theory regarding the absorption co-efficients of stellar matter, based on quantum mechanics.

2. Application of the quantum theory for analysis of the possible states of matter, and the formulation of the equation of state of stellar matter, which is designated either as normal or degenerate electron gas according to density and temperature.

3. A quantitative theory concerning the generation of energy in the stars.

In the introduction to his book *Structure and Evolution of the Stars** Martin Schwarzschild, the son of Karl Schwarzschild, writes: 'A little more than a decade ago research on the stellar interior underwent a profound change. The central cause of this change was the introduction of nuclear physics into astronomy. Nuclear physics has provided the theory of the stellar interior with the last—but not the least—of the fundamental physical processes which determine stellar structure and evolution. Thus a new and far-reaching development in this field became possible.

'Simultaneously with this new theoretical development occurred an equally far-reaching upsurge in the relevant fields of

*Princeton University Press, 1958.

observational astronomy, an upsurge due largely to the introduction of new spectrographic and photoelectric techniques.
The combination of these developments suddenly opened up an
unprecedentedly wide front of contact between observation and
theory in the research field of the stellar interior, in striking
contrast with the situation twenty-five years ago, when there was
only one major point of contact, the mass–luminosity relation.'

Perhaps, at this point, one should mention the part which
electronic computers have played since their introduction into
this kind of work in about 1950. With their aid it was possible
to work out in a very short time numerous types of stellar
models according to the various basic assumptions, and to compare these with the results of observation.

If the idea that the generation of energy in the Sun and other
stars of the main sequence results from nuclear processes, i.e.
the fusion of hydrogen atoms to form helium, is correct, then
it follows that, in the course of time, the chemical composition
of a star undergoes a change. The hydrogen content diminishes
and, at the same time, the interior becomes less and less homogeneous, since the fusion of hydrogen atoms into helium can
take place only in the vicinity of the centre, where the temperature is sufficiently high. Consequently this region is the first to
become depleted in hydrogen, so that, with the formation of
central helium cores and rising temperatures, other nuclear
processes will set in.

By combining what the nuclear physicists have to tell us about
the generation of energy with the mass–luminosity relation, the
ambiguity concerning the hydrogen content of stellar matter—
either about 30% or else 99% in the case of the Sun—is resolved. However, since the proportion of helium present in
stellar matter now enters the calculations, the theory cannot
commit itself further than to say that hydrogen and helium together account for more than 95% of a star's matter; the exact
ratio in which they stand remains uncertain. Assuming that the
relative abundance of the heavier atoms in the interior is about
the same as that of the star's atmosphere, the composition of
which can be ascertained from analysis of the spectrum, the

values appropriate to the observed metallic content can be determined to a fairly reasonable degree of accuracy. In the Sun the proportions of hydrogen : helium : metals $= 70 : 26 : 4$.

An important point arises in this connection, namely that there is a certain amount of latitude in the assumed proportion of heavy atoms. A ratio of $1 : 10$ has but little effect on the range of temperatures, densities and pressures in the interior. The theoretical central temperatures lie between 15 and 17 million degrees, and the densities between 127 and 132; the differences are thus no more than a few per cent.

The inclusion in the deliberations of models which are not homogeneous led to an understanding of the giant stars, whose very large diameters had, up to then, posed certain theoretical difficulties; they did not fit the empiric mass–luminosity relation designed for the main sequence stars. According to the new theory, however, they were regarded as stars with isothermic helium cores, in which helium, at temperatures ranging between 40 and 50 million degrees, forms the reactor fuel.

On the other hand, information provided by quantum physics in respect of the equation of state of degenerate matter has made it possible to theorise about the white dwarfs with authenticated densities of 100,000 or more, so that nowadays we have plausible models for this part of the Hertzsprung–Russell Diagram.

This phase of the theory concerning the internal structure of the stars culminated in about 1957 in a rewording of the main principle. Instead of the statement: 'The state of a star is determined by its mass and chemical composition', we now have 'The *present* state of a star is determined by its mass, its *original* chemical composition and its age'.

This summary by Martin Schwarzschild represents the key to the riddles of cosmogony, and to the understanding of the Hertzsprung–Russell Diagram as a graph of stellar evolution.

9 · Interstellar Matter

In the universe, matter also exists in other forms, apart from the stars. We can see extensive clouds of luminescence, and small bright nebulæ, which led W. Herschel to suggest that there was probably more matter in the cosmos than was then realised. The question is whether, quite apart from this directly observable matter, there is also a scattering of non-luminous matter, which will betray its presence in some way other than visible radiations. The existence of such dark matter can be detected through its effect on the motion of the stars, and on the light of the stars which lie beyond it. Diffuse matter between the stars acts as a resistance, and also exercises a gravitational effect, so that, by observing both kinds of effect in the movements of celestial bodies, it is possible to obtain some idea of the mass and density of this interstellar matter.

About a century and a half ago, Olbers suggested that light from the stars might be weakened as a result of some sort of interstellar medium. This was in an effort to get round the fatal conclusion that the entire heavens would shine with the brilliance of the Sun, if the universe, to its farthest reaches, were occupied with shining stars and the space between them were an absolute void.

The question of diffuse matter existing between the stars, and the part that this plays in the absorption of starlight, became one of prime importance with the advent of photometric methods for determining distances, since these depend on the

principle that the intensity of light diminishes in inverse proportion to the square of the distance of the light source.
Any additional weakening of the light of a star on its journey
through space, if not taken into account in calculating distance,
would make the result too great. The possibility of absorption
must, therefore, also affect the results of such investigations of
the structure of the galactic system which depend on the enu-
meration of stars according to their apparent magnitudes. The
concepts of the galactic system which Kapteyn, Seeliger,
Schwarzschild and Charlier had developed, were based on the
assumption that any interstellar absorption was very small.

Even Shapley had held to this assumption when first developing
the scale of distances for his Greater Galactic system. One of
the factors which entered into the considerations was that the
degree of weakening suffered by light would vary according to
its colour (blue light would be more affected than red) and that,
in any case, a colour change should accompany the weakening,
rather like a distant light source on Earth, which will appear
reddish through haze. The fact that distant globular clusters
appear to be the same colour as those which lie closer to us
seemed to Shapley an argument in favour of the extreme trans-
parency of interstellar space.

This argument, which is based on the assumption that
absorption and reddening go hand in hand, becomes less convincing when one takes into account the fact that absorption
also contains a neutral ingredient independent of wavelength.
It thus became a matter of some importance to try to determine
through observation the laws which govern interstellar absorption, and to relate these to the physical processes which cause it;
and in this way to learn something of the nature of interstellar
matter, and the distribution of the diffuse matter through space.

By 1920 the information regarding interstellar matter, which
had been accumulated through observation, was roughly this:

1. There exist bright irregularly shaped nebulæ, either more or
less compact, of the same kind as the Orion Nebula; embedded
in these nebulæ there are individual stars. Spectra show not only

emission lines of hydrogen and helium, but also some which cannot be ascribed to a known element. (The interpretation by Bowen that this hypothetical element Nebulium might really be ionised oxygen and nitrogen is dated 1928.)

2. Apart from the luminous nebular clouds, dark nebulæ also exist, and in our Galaxy in particular they can be seen silhouetted against the background of faint stars. By virtue of their powers of absorption, they create the impression of apparent 'stellar voids', regions in which the number of stars of a given magnitude is less than in the surrounding areas. By comparing the number of stars in these 'voids' with that in 'saturated' areas, Max Wolf developed a method of deducing the distances and densities of the dark clouds. During the years that followed this method was used extensively to investigate large areas of the Milky Way system.

3. There seems to be some connection between the dark and the luminous clouds. Obviously we are here dealing with fairly large clouds of diffuse matter, which is made to glow in the vicinity of the stars which are embedded in it. The light could be the result of radiations which stem from the nebular matter itself through the stimulus of these stars; alternatively, it could be merely reflected starlight. Analysis of the spectrum supplies the answer: the dark clouds consist, at least partially, of solid particles, cosmic dust.

4. In 1904 J. Hartmann, while carrying out an investigation of the spectroscopic binary Delta Orionis, noticed that the Fraunhofer line K, which is ascribed to calcium, fails to participate in the periodic shift of the spectral lines which is due to orbital motion. In addition, this line is sharper and finer than all the other lines which tend to be broad and not very well defined. Two explanations were possible: either this binary system is surrounded by an extensive cloud of calcium gas, or the light from these two stars just happens to pass through an isolated cloud of calcium gas.

Such 'stationary' or 'interstellar' calcium lines were subsequently also discovered in other stellar spectra, and in 1919 came the discovery that the D line of sodium also occurs as an interstellar line. In this way it was established that free gas atoms are present in interstellar space, and that these cause absorption lines in the spectra of stars. Whether these gases are concentrated into individual clouds or are more or less evenly distributed through space was still very much a matter of conjecture at that time.

From these clues about the interstellar medium were born the problems which occupied research during the years that followed. They were aimed principally towards two objectives. The first was the quantitative evaluation of the effect of interstellar absorption on the study of the structure of the galactic system, and the extra-galactic universe; the second was to find out more about the nature of the interstellar medium, in order to learn what part it plays in the processes of the universe, in the evolution of stars and stellar systems.

Locating the dark clouds and determining to what extent the light from stars is dimmed in its passage through them are factors which must be known before one can answer the question of whether or not the apparent irregular distribution of stars in the Galaxy points to a correspondingly uneven distribution through space. The way the dark clouds are concentrated in the Milky Way explains the apparent distribution of the spiral nebulæ symmetrically to the Galaxy; the spirals seem to be concentrated towards the galactic poles, while in the Milky Way itself they are absent.

In the controversy concerning the 'island universe' nature of the spiral nebulæ, this apparent alignment with the Milky Way was thought to be a cogent argument for the belief that the spirals might be parts of the Milky Way system. But the discovery of some very faint spirals within the Milky Way itself confirmed the idea that the distribution of dark matter in our Galaxy distorts our impression of the extragalactic universe. Generally speaking, our view in the direction of the galactic plane fails to penetrate beyond the bounds of our own Galaxy. Only in some places, where successive layers of dark clouds happen to leave a gap, are we able to see, as through a window, into the farther reaches of space beyond the confines of our Galaxy.

The question of general absorption and its effect on the apparent distribution of stars and star systems was clarified in about 1930, insofar as it was possible to reach the following conclusions:

1. The matter responsible for the general absorption is contained in a layer parallel to the galactic plane whose thickness is in the order of about 500 parsecs (1,500 light-years). The average absorption is approximately 0·6 magnitude per kiloparsec, but varies locally from 0·5 to 4 magnitudes, an indication that interstellar matter is not uniformly distributed throughout the layer; instead, the dark clouds are ranged rather like a series of separate curtains one behind the other.

2. The absorption is selective, that is to say, not the same for all colours; in particular it tends to be more pronounced for the photographic wavelengths than for the visual. Thus starlight not only undergoes a general weakening, but also reddening equivalent to several tenths in the colour index. The change in the relative energy distribution of the continuous spectrum of the stars as a result of selective absorption brings about an apparent lowering of the colour temperature. If the true colour temperature of a star can be determined from the character of the spectrum, i.e. the linear intensities, then the degree of reddening and hence also the degree of absorption itself can be found.

3. From the observed absorption law it can be deduced that absorption results from scattering due to solid particles; thus, the constituent of interstellar matter which is responsible for general absorption is cosmic dust. A number of theories exist as to the size of these dust particles, but observational data remain inconclusive. The density of the dust content is estimated to be in the order of approximately one ten-thousandth part of the mass of the Sun per cubic parsec; if all the matter concentrated in the stars were spread uniformly through the entire Galaxy, it would be a thousand times more dense than the cosmic dust.

Consequently the dust represents no more than an infinitesimally small proportion of the total mass of our stellar system.

The problem of the interstellar gas, whose only known components in 1930 were the atoms of calcium and sodium, seemed at the time to be the following: From observation of the stationary lines of calcium in the spectra of numerous stars one can assume a wide distribution of calcium and also sodium, without entering into the question of whether this distribution is uniform, or whether it is similar to that of the cosmic dust. Measurements of radial velocities show clearly that interstellar calcium and sodium share the general rotation of the Galaxy.

The strengths of the sodium and calcium lines increase fairly steadily with the distances of the stars in whose spectra they occur. This fact tends to demonstrate that, on the whole, the absorbing medium is spread reasonably evenly through the Galaxy, and so presents the possibility of determining the distances of stars by means of the strength of the interstellar lines and these distances will not be distorted by general interstellar absorption.

The Fraunhofer lines H and K emanate from calcium atoms which have lost one electron. This now raises the question of how the ionisation of atoms can take place in interstellar space. In the discussion about the states of matter in the interiors of the stars, the idea was put forward that radiation may be regarded as having the characteristics of a gas in which radiation quanta represent the molecules. We spoke of a 'photon gas' which can be described in terms of density, pressure and temperature, just as any normal gas. When this photon gas leaves a star and spreads through space, it naturally becomes more and more rarefied; the process may be expressed as the reduction in the intensity of radiation in inverse proportion to the square of the distance. The number of photons per cubic centimetre, i.e. the density of energy, diminishes whilst the energy of the individual packets remains unaltered, so long as there is no collision with any material particle, which would entail a transfer of some or all the energy to the latter.

111

The radiation emitted by the stars of the Galaxy in all directions fills the space between the stars with a photon gas, and there is a mutual reaction with any matter which is also present. However, in contrast to the interiors of the stars, the density of this mixture of dust particles, gas atoms and photons is so extraordinarily low that collisions between the components are extremely rare. An individual photon can travel for years, and cover distances which are equivalent to galactic dimensions, before it meets a dust particle or an atom to which it can transfer its energy.

The sum of the energy transferred through collisions, in effect the number of such collisions, determines the temperature of a gas. The infrequency of collisions with photons, which material particles in interstellar space suffer as a result of the low energy density, means that the temperature of such matter is raised to no more than 3° above absolute zero.

A collision between a photon and an atom presents a completely different picture. The energy which the photon carries with it on its unimpeded way through space until impact is determined by the effective temperature of the star from whose atmosphere the photon originated. The interstellar photon gas thus contains particles whose energy corresponds to temperatures between 2,500° and 25,000°, according to the spectral type of the star of origin. When a collision with an atom does occur, then this happens with great violence, and the photon has sufficient energy to deprive the atom of one of its electrons. In this way singly, doubly or multiply ionised atoms can be produced in the interstellar gas in proportion to the energy distribution of the radiation.

To sum this up, we may say that the amount of radiation filling interstellar space, that is to say the energy density, decides the temperature of the matter, and thus its radiation; the quality of radiation, that is to say the relative energy distribution, determines the degree of ionisation, and with it the lines, which can show up as interstellar absorption lines. The inherent temperature of cosmic dust lies at around 3°K, if one takes into account only the energy density of radiation; on the other hand, the degree of ionisation of the interstellar gas is approximately equivalent to a temperature of 10,000–15,000°.

In the ten years from 1930 to 1940 a number of advances were made in the study of the interstellar medium. Accurate analysis of the interstellar lines H and K of calcium in spectra with high resolution revealed details which could only be explained by the assumption that on its journey to us the light from the star in question had to pass through several successive clouds of inter-stellar material, each moving at a different speed. This is rather reminiscent of the arrangement of the dark clouds in the manner of a series of curtains, as deduced from the general absorption and reddening of starlight. It seems to indicate some sort of connection between dust and gas.

As well as the lines of calcium and sodium, numerous other absorption lines of interstellar origin were subsequently dis-covered. By 1942 no fewer than thirty-five such lines had been listed, which, apart from eight unidentifiable ones, were attri-buted to the elements potassium, sodium, calcium, iron, titanium and compounds of carbon and hydrogen or nitrogen. The application of the theory of thermal ionisation, which was developed for stellar atmospheres, to this problem of the inter-stellar medium permitted a full analysis of its physical state.

The result of this analysis is: The mean density of this matter is equivalent to one quadrillionth of the density of water, approximately the same as the mass of one hydrogen atom per cubic centimetre. A volume of one cubic metre con-tains 15 million hydrogen atoms, and besides $14\frac{1}{2}$ million free electrons no more than 111 atoms of sodium, 15 of potassium, 6 of calcium, and only in a volume fifteen times as great is one likely to come across one atom of titanium.

At about the same time, analysis of the composition of the atmospheres of the Sun and several other stars led to the realisa-tion that hydrogen and helium account for most of their masses, and that all the heavier elements together amount to barely 3%. The conclusion was reached that even the interiors of the stars consist almost exclusively of hydrogen and helium, and, further, that in the universe as a whole matter exists principally in the form of the lightest atoms.

While these empirical and theoretical investigations were being made into the nature of the interstellar medium, whose presence can be detected through the absorption of starlight, efforts were being directed towards discovering the exact connection between the luminous nebulæ and stars. If individual stars are responsible for the luminosity of the nebulæ, irrespective of whether they excite the gases in the nebulæ to produce the emission lines, or whether starlight is merely reflected from solid particles, then a corresponding distribution of luminosity should be noticeable in the nebula.

If the nebular matter is absolutely evenly distributed, we should see the nebulæ as circular discs; the brighter the star embedded in it, the greater the diameter. This is indeed the case for a particular type of nebulæ; because these resemble the disc-like appearance of the planets, they are known as 'planetary nebulæ'. The Ring Nebula in Lyra (M.57) is one of this type and is a most impressive sight through a telescope.

But even in instances where this ideal shape is, to a greater or lesser degree, blurred over because of the structure of the nebular material, as in the Veil Nebula in Cygnus, which seems to be composed of fine filaments, it can almost always be shown that a star is present. Regardless of whether we are dealing with an emission nebula, or a reflection nebula, the apparent extent of the area of luminosity always increases with the brilliance of the star which stimulates it.

Emission nebulæ and reflection nebulæ differ by the sort of stars which are associated with them. Genuine emission nebulæ are always associated with stars of very high temperatures; the stars embedded in the Orion Nebula as well as the nuclei of planetary nebulæ are B- or even O-stars, with effective temperatures well in excess of 20,000°. Clearly, the nebular matter is made to shine as a result of highly energetic radiations from these very 'hot' stars.

In contrast to emission nebulæ, which have their own spectral characteristics, the spectra of reflection nebulæ resemble those

114

of the illuminating bodies. Therefore, although the two types of nebulæ may not differ essentially in composition and structure, their spectral characteristics are determined exclusively through the nature of the stars which are embedded in this diffuse matter.

In this connection, there arose another question, which began to be debated with the flare up of the nova in Perseus in the year 1901. In the vicinity of this star nebular material was observed, and this in time seemed to extend in all directions. Did the occurrence occasion actual movement of gaseous masses, expelled from the nova during the flare up, or, alternatively, could this phenomenon be the result of a wave of light which was emitted by the star and could be seen spreading through a normally dark cloud of matter surrounding the star? A decision in favour of the first of these suggestions was only reached some twenty years later, after further novæ, principally the bright Nova Aquilæ of 1918, and Nova Pictoris of 1925, had provided data for spectral analysis and for computing the real expansion velocity of the nebular shroud.

J. Hartmann's epic telegram from La Plata, where he had observed Nova Pictoris, to the editor of *Astronomische Nachrichten* in Kiel—'Nova problem solved star expands and bursts' marks a turning point; from then on the arguments in favour of a genetic–causal relationship between the 'new' stars and the gaseous nebulæ, between the stars and interstellar matter in general, began to increase in number. The Crab Nebula can be regarded as the relic of a supernova, and from the rate at which matter is expanding—in the order of 1,100 km./sec.—it is possible to estimate the date of the outburst; it must have been seen from the Earth in the second half of the tenth century. Chinese records tell us of the appearance of a 'new star', brighter than the planet Venus, in the year 1054, and in roughly the same region of the heavens where the Crab Nebula is to be found.

Largely as a result of improved methods of observation, the last decade or so has brought new insight into the nature and distribution of interstellar matter. The discovery in 1951 of

115

interstellar hydrogen as the source of ultra-shortwave radiation with a wavelength of 21 cm. became the basis for a comprehensive study of the distribution of hydrogen gas in the Galaxy. Through this we have learnt that interstellar hydrogen mixed with cosmic dust is concentrated in the main plane of the Galaxy along the spiral arms. Measurements of the degree of polarisation which light undergoes in its passage through the interstellar medium, that is to say, the change in the direction of oscillation, complemented the photometric studies, which were concerned with absorption and reddening; these observations led to the discovery that the particles of interstellar dust are adjusted parallel to the plane of the Galaxy, and thus indicated the existence and strength of interstellar magnetic fields, which cause elongated particles to align themselves in much the same way as iron filings will do along the lines of force in the case of a magnet.

As well as such studies of our own stellar system, similar investigations were undertaken with regard to the Andromeda Spiral and also other extragalactic systems. These provided data of the amount and distribution of diffuse matter dependent upon the type of system. And then, in more recent years, there has been increasing evidence that also the space between the galaxies is not entirely devoid of matter, and that there is such a thing as intergalactic matter. It will be obvious from this what a wealth of problems faced the theorists, who were expected to answer such questions as what mutual effects the stars and the diffuse matter have exercised on each other since their origin; what effects have the magnetic fields, and what is the general rôle played by this scattered material in cosmic events, in the birth and extinction of stars.

Let me, at this stage, sum up the evidence which has so far been accumulated regarding the question of interstellar matter.

Interstellar matter makes up at most one tenth of the entire mass of the Galaxy; the rest is concentrated in the stars. The diffuse matter is a mixture of dust and gas consisting of about 1% of the former and 99% of the latter.

The chemical composition may be expressed proportionally as hydrogen : helium : heavy elements = 60 : 38 : 2. On the physical side, about 10% of the gaseous content is ionised at a temperature of roughly 10,000°, and 90% is neutral at a temperature of 50°, though this may rise locally to several thousand degrees for short periods.

10 · Birth and Death

Henri Poincaré in the preamble to his *Lectures on the Hypotheses of Cosmogony*, which he gave at the Sorbonne, Paris, in 1910–11, stated that: 'It is not possible to behold the spectacle of the star-strewn heavens without wondering how it came into being. It might perhaps be wiser to wait until such time as we have gathered the last remaining shred of evidence before venturing an opinion on this matter. However, if mankind were in fact so exemplary, if we were simply curious without also being impatient, the sciences would never have developed, and we should be content to do no more than simply live out our lives. It is in our natures, however, to suggest solutions before the time is ripe, armed with but a few of the barest possible facts. This is the reason why cosmogonistic hypotheses are so numerous and so varied, and why new theories come out almost daily, and are for the most part just as uncertain and as plausible as those which they seek to replace. Eventually they simply take their place alongside the older theories without demolishing them.'

In the introduction to the section on cosmogony in the *Encyclopædia of the Mathematical Sciences*, I raised the following points in the year 1933:

'Cosmogony is a branch of mathematics in which the efforts of outsiders and visionaries outweigh those of the so-called experts both in their scope and in quantity. Just as the idea of the "*perpetuum mobile*" nags the inventive mind, and new solu-

118

tions to the riddle of gravitation make their appearances annually, so the question of the origin and destiny of the universe is constantly being "solved". And even though almost all of the theories, which have been put forward at some time or another, have eventually been replaced or resolved, they still manage to crop up every so often even today—sometimes in their original forms apart from one or two minor modifications, and sometimes with more radical changes. One can hardly expect things to be otherwise, since there are so few definite facts to go on, and the whole subject is founded almost exclusively on hypothesis.'

Until the turn of this century, cosmogony was principally concerned with the planetary system; it was unthinkable that the harmony, which could be seen to exist, should be a matter of pure chance. Poincaré put it this way: 'One could well suppose that an infinite intelligence had from the outset ordained the order of things for all eternity, and indeed this explanation was considered satisfactory enough in earlier times. Today, however, the matter is no longer considered quite so simple. If the order which we observe around us is not the result of chance, and we are not inclined to attribute it to the outright creation of a supernatural power, then it must have developed out of chaos; in other words, the stars must have evolved.'

The idea of the development of order out of disorder has dominated cosmogony since Kant suggested it in 1756 in his treatise *Allgemeine Naturgeschichte und Theorie des Himmels*. Kant supposed the beginning to have been complete chaos; Laplace postulated a primæval sun, a rotating ball of gas, which in itself was already a sign of some sort of order, although he failed to account for the origin of this first stage in the creation of order. In fact, this did not come until almost the end of the century, when Norman Lockyer suggested that the stars began as condensations out of clouds of cosmic dust. The problem of the origin and evolution of the stars stems from this idea and has become one of the most important considerations of the day.

In conjunction with Hertzsprung's discovery that some of the stars are giants and some are dwarfs, Lockyer's concept of 'ascending' and 'descending' stages of stellar evolution was incorporated into the pattern of H. N. Russell's *Giant and Dwarf Theory of Stellar Evolution*, which began to predominate from about 1913 on. According to the Giant–Dwarf theory, red giants represent the first stage in the evolution of stars. A red giant is a ball of gas with low density and temperature; as a result of contraction it becomes hotter, and passes through the yellow giant stage before reaching its peak in the form of a white star of great temperature and luminosity. Since further contraction fails to provide sufficient energy to make up for that lost through radiation, the temperature and luminosity fall off comparatively rapidly. So the star begins what one might call the return journey, passing through the stage of our own Sun until it becomes a red dwarf, which, owing to its diminishing luminosity, gradually fades from sight.

The life-cycles of the stars portrayed in the Hertzsprung–Russell Diagram proceed according to this idea from the top right horizontally across to along the branch of the giants until they meet the main sequence which runs diagonally from the top left to the bottom right of this diagram; the cycle now follows the main sequence. In the twenties it was generally assumed that this sort of development was bound to follow from Eddington's theory of the internal structure of the stars. In reality, of course, this theory can only tell us something concerning the particular state of stellar matter, but nothing about the changes of state in the course of time. If one wishes to learn something about the way in which a star is developing, it is necessary to discover how the star generates the energy which it radiates.

If the Sun, as Helmholtz's contraction theory would have it, had condensed to its present radius out of a widely spread mass of matter of very low density, then it would have gained gravitational energy in the process. More than half of this energy would have been stored in the interior, while the rest would have been lost through radiation during the evolutionary stages. Calculated on this basis, the Sun would appear to be at the most some 23 million years old. Geological and biological evidence, however, gives the age of the Earth, and thus also the time during which the Sun has been shining with undiminished brilliance, as several thousand million years, so that the time scale suggested by the contraction theory is about a hundred times smaller than it ought to be.

The inescapable conclusion must be that only atomic energy sources can explain the Sun's ability to emit radiations over a period of approximately several thousand million years. In about 1900, radioactive decay was thought to be the energy source, but simple calculation soon showed that, even if the Sun consisted entirely of uranium, it could not continue to radiate energy at the rate at which it does so for more than a few million years.

It was not until Einstein equated mass and energy, which included the possibility of matter being converted into energy—in accordance with the equation: Energy = Mass × square of velocity of light—that one began to arrive at the correct order of magnitude. When a lump of coal is burned in the normal way, one kilogramme will produce, at the most, 8,000 calories, but if the entire mass were converted into energy it would yield thousands of millions of times this value, 22 billion calories. Thus a Sun made up entirely of coal would burn away completely in 6,000 years, whereas the conversion of only one-thousandth part of the Sun's mass into energy could cover its radiation for at least 10,000 million years.

So long as little or nothing was known about the processes in the interiors of the stars by which atomic energy was released, it was possible to interpret Einstein's equation in a number of different ways; according to how much of the mass was presumed to be turning into radiation, various cosmogonistic time scales were possible. The age of the Sun, its past duration, depends on its original mass, and on the productivity of the atomic energy sources; the total dispersal of its present mass through radiation would determine the maximum extent of its future. To the 'short' time scale of Helmholtz's contraction theory, in which the reckoning is in millions of years, there was added a sort of 'intermediate' scale reckoning in thousands of millions, even if only a small fraction of the total mass, something in the order of 1/1,000 could be set free in the form of atomic energy; the 'long' time scale, on the other hand, talks of the past in terms of millions of millions of years with a considerable proportion of the total mass having been converted into energy and so lost.

The mass–luminosity relation, which arose from hypotheses concerning the internal structure of the stars and was subsequently confirmed through observation, meant that so far as the giant–dwarf theory of stellar evolution is concerned development downwards along the main sequence must be accompanied by a definite decrease in stellar mass. The question of whether this loss of mass is the result of emission of actual material particles, or whether the mass is converted into energy and so leaves the star as radiation remained open. In the decade between 1920 and 1930 the idea of 'annihilation of matter' played an important rôle in such cosmogonistic arguments as regarded the giant–dwarf evolution pattern as a scientific fact, and the H–R diagram as the corresponding graph.

The true solution to the problem began to emerge in 1929, when Atkinson and Houtermans, working on some suggestions put forward by Gamow, showed that in the interiors of stars, at temperatures of around 20 million degrees, nuclear processes are able to take place which result in the formation of higher elements from the known fundamental particles of matter—at that time protons (nuclei of hydrogen) and electrons. The mass of the combined nucleus is less than the sum of the masses of its individual components, the difference appearing in the form of energy. The most prolific process is the combination—the term nowadays is 'fusion'—of four hydrogen nuclei to make one helium nucleus. The deficit in mass in this case amounts to 0·8%. If the Sun had originally consisted entirely of hydrogen, then its radiation over the past ten thousand million years would entail the conversion of only 12% of this hydrogen into helium.

In 1932, on the evidence of the mass–luminosity relation, Bengt Strömgren put forward the suggestion that the colour–magnitude diagrams of the stellar clusters could also be interpreted as evolutionary diagrams of groups of stars of the same age with the same hydrogen content; it was on this point upon which the giant–dwarf hypothesis as given by Russell finally foundered. Because the lines of equal hydrogen content in the colour–magnitude diagram run parallel with the main sequence, and because the farther they lie to the right, the lower the hydrogen content, it became apparent that stellar evolution as shown on the H–R diagram (assuming that the source of energy is hydrogen fusion) does not follow the line of the main sequence in the direction of diminishing luminosity and temperature; instead the evolutionary path runs at right angles to the main sequence in the direction of slightly rising luminosities and falling temperatures.

It is interesting to recall that in 1921 at the meeting of the Astronomical Society in Potsdam, the first after World War I, Wiechert, the well-known geophysicist from Göttingen, put forward the theory that in the course of their evolution the stars actually gain in mass, and that consequently their evolutionary paths on the H–R diagram should, in fact, run in the opposite direction to that indicated by Russell's giant–dwarf theory. At the time—I well remember this dramatic episode at the meeting—we all ridiculed the old man's ideas. We were still unaware of the precarious footing on which the universally accepted giant–dwarf theory really stood.

The picture became clearer from about 1939 on, when Bethe and von Weizsäcker were able to explain the processes which were likely to be the sources of solar and stellar energy. The atomic reactor inside the stars could be understood and evaluated mathematically in the light of developments in theoretical and experimental nuclear physics. In this way essential data for calculating stellar models and for reconstructing the life-cycles of stars could be obtained. From the 'balance sheet' of energy, the upper limits for the ages of stars at present to be found in the Galaxy could be deduced. It was possible to conclude that the B type stars having the highest temperatures and luminosities were hardly likely to be older than 50 million years, while the maximum age of the Sun could be as much as 35,000 million years, if all the helium which it now contains were produced in its fusion reactor out of hydrogen.

Since, at that time, the 'age of the universe' based on the hypothesis of an expanding cosmos was still considered to be in the order of 3,500 million years, and since the Sun could certainly not be older than the universe as a whole, the inevitable conclusion had to be reached that only 10% of the helium now present in the Sun could have been created as a result of hydrogen fusion within the Sun itself; by far the greater proportion, some nine-tenths, the Sun must have inherited at birth. The same sort of argument was also valid so far as the higher elements were concerned, for whose composition quite different temperature conditions are required than could ever have obtained in the interior of the Sun.

Once the implications of this theory had been worked out in greater detail, it was realised that ideas concerning the age sequence of the stars required complete revision. The red giants no longer stood at the beginning of the ascending branch of stellar evolution; neither did the red dwarfs mark the end. It seemed far more likely that the red giants represented the oldest known stars, older than the stars of the main sequence of which the Sun is a member. They were stars in a stage of evolution in which the hydrogen fuel stores have largely been exhausted, and at whose immediate centre a core consisting wholly of helium has been formed; here, at temperatures ranging between 100 million and 250 million degrees, heavier nuclei were being built up.

The relative frequency of the elements observable today raises the question of when and where the construction of the higher elements from fundamental particles (protons and neutrons) took place. In about 1945 there still seemed no way of progressing beyond the lightest nuclei from hydrogen by way of helium in the interiors of the stars, and obtaining heavy nuclei in the frequency in which they exist in the stars. It was therefore proposed that the elements were created at temperatures of millions of degrees within the primæval universal mass, whose explosion gave rise to the expanding universe, and that, as from this moment, their frequency has remained virtually 'frozen'.

Nuclear physical considerations, together with the results of astrophysical observations, gradually undermined this concept. Its place was taken by the view that originally the universe must have consisted of pure hydrogen, and that the heavy elements developed only in the course of evolution in the interiors of the stars at extremely high temperatures and pressures; to some extent they may also have been, and still be, created through explosions like those of the supernovæ. This hypothesis is supported by the discovery of stars with considerably smaller amounts of heavy elements than the Sun and the stars of the main sequence, as well as the constitution of interstellar matter. Among the stars of Population II, which form the 'halo' of our Galaxy, there are some with a metallic content which is less than one part per thousand. Perhaps such stars are evidence of the composition of matter as it originally existed during the cataclysmic early phases in the evolution of the Galaxy, when the first stars and globular clusters were being formed.

More exact data as to the age and evolutionary stage of the stars and star systems was reached by examining the colour–magnitude diagrams of galactic clusters and globular star clusters. The calculation of theoretical stellar models showed that a star of the main sequence, in whose interior the hydrogen fusion reactor is operative, changes its condition, as defined through luminosity and effective temperature, only gradually; it retains its position in the main sequence on the colour–magnitude diagram until the depreciation in hydrogen becomes quite noticeable, and a core consisting predominantly of helium has been formed. The loss of hydrogen is accompanied by displacement of the star from the main sequence, slowly at first, in the direction of diminishing effective temperature; then, when the central temperature has risen sufficiently, and the helium reactor has become active, it finally moves in the direction of increasing luminosity with further reductions in the effective temperature, and into the giant branch.

The greater the mass and the luminosity of a star, the earlier it leaves the main sequence, taking a group of stars of the same age. If the star clusters are arranged according to the characteristics of their colour–magnitude graphs in such an order that the series is determined by the location of the upper end of the main sequence, then such an order will also be indicative of age. Star clusters whose main sequences reach farther to the top, and which still hold hot B-type stars with luminosities of 10,000, are the youngest; the shorter the main sequence, the older is the cluster.

Thus the age of the double cluster in Perseus is estimated to be only 4 million years, that of the Pleiades 80 million, the Hyades almost 1,000 million, and the cluster Messier 67 over 4,500 million years. The globular clusters which, in the framework of the old giant–dwarf theory of stellar evolution, were regarded as relatively young because they contained red giants, now proved to be the oldest objects in our Galaxy; aged between 6,000 million to 10,000 million years, they are probably as old as the Galaxy itself.

Generally speaking, so far as the star clusters are concerned, we have information only about the upper part of the colour–magnitude diagram, the part which contains stars of the highest absolute magnitude; the apparent limit towards the weaker stars is set by observational difficulties. There is therefore some uncertainty as to the overall number of bodies belonging to a particular star cluster, as well as the location of the weaker stars in the colour–magnitude diagram relative to the main sequence. In a few instances it has, however, proved possible to speculate even about the lower portion of the diagram, and this has led to understanding of the earliest phases of stellar evolution.

One such factor is the existence of what are known as 'stellar associations'; these were first pointed out by the leading Soviet astrophysicist Ambartsumian. They consist of loose groups of on the whole not very numerous, extremely hot stars, often embedded in luminous nebulæ, such as the stars of the well-known Trapezium in the Orion Nebula. The existence of these associations, which by virtue of their loose structure do not stand out from the general star field as concentrations in the same way as

actual star clusters, was initially received with the same doubt
as Ambartsumian's opinion that the associations contain very
young stars which have formed out of individual condensations
in clouds of interstellar matter.

Ever since the Congress of the International Astronomical
Union in Rome, when the matter was thoroughly discussed, the
associations have been accepted as definite entities in the structure of the universe, and their status as representative of the
earliest stages of stellar evolution has been recognised. With regard to the colour–magnitude diagrams of the stellar associations, the brightest, that is to say hottest, stars of great mass lie
on the main sequence; their internal atomic reactors are operative. The fainter stars of smaller mass, on the other hand, lie
beside the main sequence in the diagram, nearer the lower
temperatures; it was this kind of star which, in the chapter about
the Hertzsprung–Russell Diagram, was said to be just as much
a sub-giant as a super-dwarf. According to more recent ideas
about the internal structure of the stars, these are stars which
are still in the contraction stage of evolution, and have not yet
reached sufficiently high internal temperatures for the hydrogen
fusion reactor to become operative. So far as the H–R diagram
is concerned, they are still on the way to becoming main
sequence stars.

Even more recently, examples of star clusters have been found
which do not contain any main sequence stars at all; all the
members of this kind of cluster are apparently on the way to
the main sequence. The ages of these very young star clusters
are hardly likely to be more than half a million years or so.
This means that the stages of stellar evolution could be traced
right back almost to their very start both in theory and by
observation; it is interesting to speculate whether, in fact, the
earliest stage also can be made available to observation.

It becomes self-evident that not all the stars could have been
formed at the same time as the Galaxy, since stars exist whose
maximum ages, even assuming the most prolific energy sources,
cannot be more than a few million years; it also follows logically
that stars are being born at this very moment. By tracing the

stages of development backwards, we find that the evolutionary path brings us to the clouds of interstellar matter as the birthplaces of stars, if we agree with Ambartsumian and regard the stellar associations as the cradles of the stars.

The differences in the chemical constitutions of the stars, in particular differences in the amounts of heavy elements present in stars of varying ages in Populations I and II, lead to the conclusion that a certain proportion of these elements was formed in the first generation of stars, which in turn formed during an early phase of the Galaxy out of a hydrogen–helium mixture, and that these heavy elements reached later generations of stars by way of the interstellar material.

The 'Gesellschaft Deutscher Naturforscher und Ärzte' instituted an open competition in 1956 for theses on *The Origin of Stars through Condensation of Diffuse Matter*. Three of the works submitted, one each from Germany, the United Kingdom, and the United States of America, tied for first place, and were published in book form in 1960. The results may be summed up as follows:

1. Throughout the entire existence of the Galaxy (approximately 10,000 million years) stars have been in continuous creation, initially at ten to twenty times the rate as at present. New stars are still being formed.

2. The creation of stars takes place in the spiral arms through nebular agglomerations of interstellar matter when gravitational attraction becomes greater than dispersing forces.

3. After the formation of a few bright O-type stars, conditions in the vicinity become more favourable for the formation of further stars.

4. The formation of single stars can be satisfactorily explained only if they are massive O-type stars, more than ten times the mass of the Sun. Normal stars of smaller masses must have originated within clusters or associations with something like a thousand times the solar mass.

5. The formation of stars seems to take place only if conditions are favourable (high density, sufficient mass, strong mag-

1. Karl Schwarzschild Observatory at Tautenburg near Jena

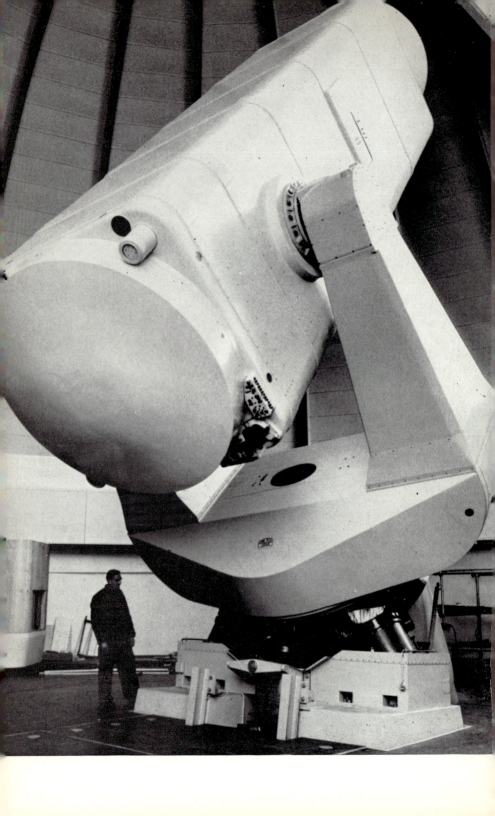

3. The Andromeda Spiral

2. The two-metre universal reflector constructed by Zeiss of Jena

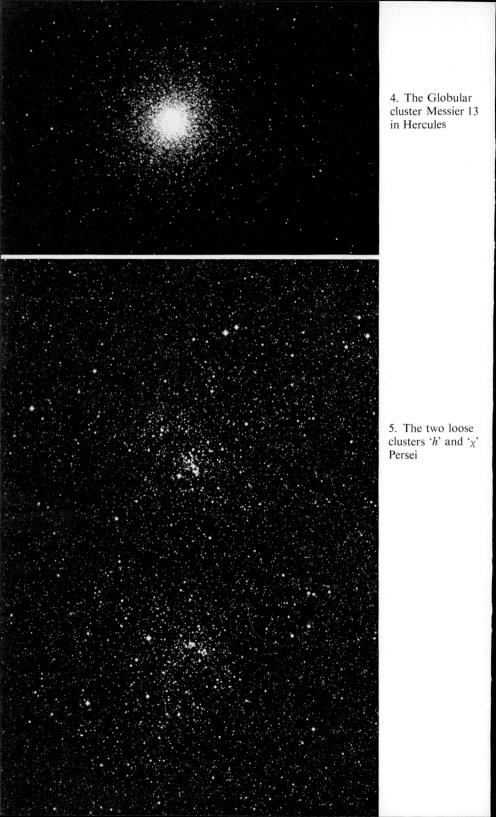

4. The Globular cluster Messier 13 in Hercules

5. The two loose clusters '*h*' and '*χ*' Persei

6. A section of the galactic field in Virgo

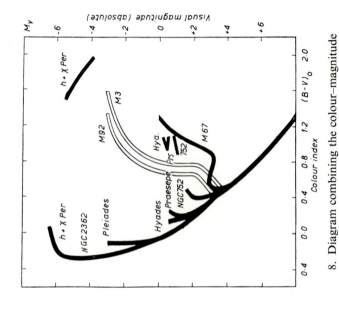

8. Diagram combining the colour–magnitude diagrams of open and globular star clusters

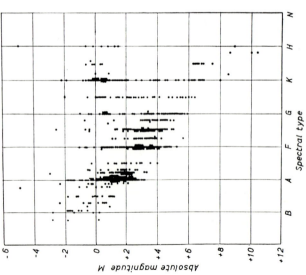

7. This version of the Hertzsprung–Russell diagram was published by H. N. Russell in 1914

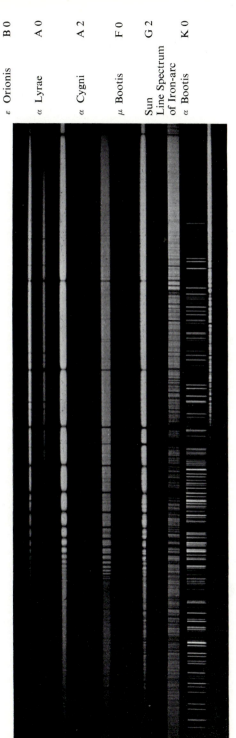

STAR	SPECTRAL TYPE
ε Orionis	B 0
α Lyrae	A 0
α Cygni	A 2
μ Bootis	F 0
Sun	G 2
Line Spectrum of Iron-arc	
α Bootis	K 0

9. Typical stellar spectra

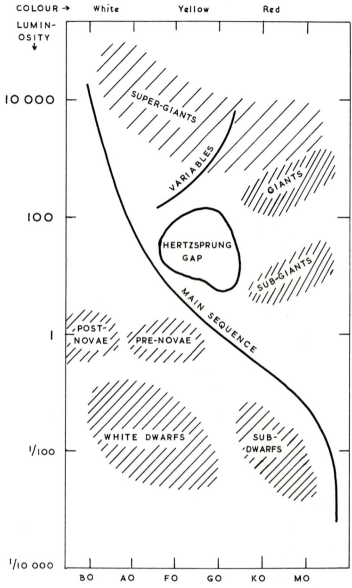

10. Schematic Colour–Magnitude diagram (1955)

11. The Pleiades—a galactic cluster

12. The Great Nebula in Orion

Appendix · Explanation of Illustrations

1 The dome housing the universal reflector at the Karl Schwarzschild Observatory at Tautenburg near Jena, Germany. (*Photo:* Tautenburg)

2 The 2-m. universal reflector which was completed in 1960 was constructed by Zeiss of Jena with the author Hans Kienle as consultant. The main mirror is spherical and has a diameter of 205 cm. and a focal length of 4 m. It can be used either in conjunction with a Schmidt correcting lens (134 cm. diameter), or with over-corrected secondary mirrors as a Cassegrain or Coudé system. The celestial photographs in this book were taken with this instrument using the Schmidt system. The Cassegrain system with a focal length of 20 m. (the light emerges through the declination axis) is used for photo-electric and spectrographic work. In the Coudé arrangement (focal length 92 m.) the light rays travel by way of a number of plane mirrors in the cross-beam of the fork mounting through the polar axis into a high dispersion spectrograph situated in the basement of the observatory. (*Photo:* Tautenburg)

3 The Andromeda Spiral is a system similar to our own Galaxy, and may be regarded as representative of the latter in almost every detail; we see this galaxy from one side. Its two companions are 'elliptical galaxies' which appear to

130

netic fields, and the minimum of internal movement). For this reason not all interstellar matter has yet been used up.

I am quite aware that I have followed only the path of stellar evolution from the purely speculative giant–dwarf theory through to present day concepts based on sound theoretical and observational foundations. The purpose has been to show how, in this particular field, our ideas have taken shape and altered in the course of the first half of this century; much is the result of a better understanding by physicists of the structure and behaviour of matter. The theme has not been exhausted by any means and a great deal more could be said regarding the origins and evolution of galaxies, and the problems of cosmogony which are concerned with the universe as a whole. Nor have I dealt with the subject with which cosmogonistic thinking during the nineteenth century was almost exclusively concerned, the origin of the planetary system. Let me therefore conclude with a brief reference to this topic.

It is fairly safe to say that our planetary system is not unique, but is likely to be a relatively common phenomenon of cosmic evolution, occurring in similar form elsewhere in the universe. The idea that the Solar System was formed as the result of the close approach of two stars—which was in favour at the beginning of the century—has now been rejected. It seems very much more probable that the Sun, the planets, and their satellites were born 'cold', individually, at more or less the same time, and formed from condensations in an extensive cloud consisting of a mixture of gas and dust.

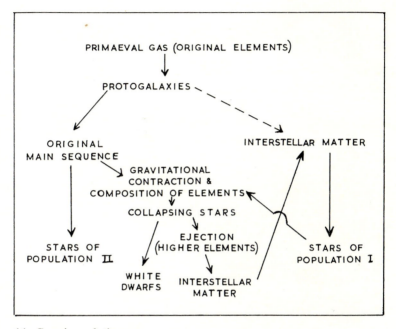

14. Cosmic evolution

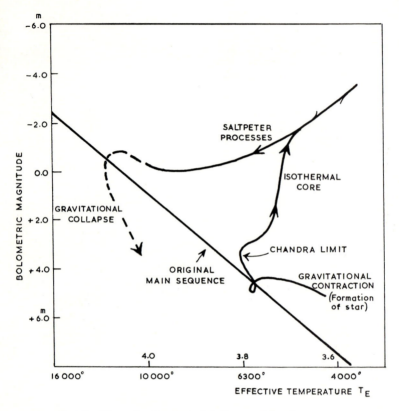

13. The life cycle of a star of the mass of the Sun

consist entirely of stars, having no spiral arms containing diffuse matter. The three galaxies shown, together with our own Galaxy, all belong to what is known as the 'local group' which consists of some twenty members of widely varying dimensions. These three objects lie at a distance of about 2·5 million light-years from us, and their masses are in the order of 160, 2·5 and 1·5 thousand million times that of the Sun. (*Photo:* Tautenburg)

4 The globular cluster Messier 13 in Hercules is one that belongs to our own Galaxy; about 150 of similar objects have been listed, though the actual total is thought to be at least double this number. In general they lie at mutual distances of about 7,000 light-years within the 'halo' whose diameter is 150,000 light-years, and their frequency decreases with distance from the centre of the system. The diameters of these objects range between 50 and 350 light-years, while the number of stars which such a system may contain is in the order of several hundred thousand. The faintest stars which can be distinguished in this photograph have absolute magnitudes brighter than the Sun. (*Photo:* Tautenburg)

5 The two loose clusters '*h*' and '*χ*' Persei are examples of 'open' or 'galactic' clusters. In contrast to the globular clusters these objects occur only in the galactic plane, and it has been estimated that there are some 15,000 of them; so far 500 have been identified. These clusters have diameters ranging between 10 and 30 light-years, and are about 300 light-years apart. The number of stars they contain can be several thousand, but on average it is in the region of 100. The galactic clusters vary in age; at something in the order of 4 million years, the double cluster in Perseus is one of the youngest, while the oldest known is probably Messier 67, which is almost 5,000 million years old. Compare this illustration with Plate 8. (*Photo:* Tautenburg)

6   A section of the galactic field in Virgo showing a number of extragalactic objects of various kinds: spirals seen from various angles (from above, obliquely and edge on), and elliptical galaxies with varying degrees of eccentricity. All the objects on this plate which are not star-like in appearance are galaxies, stellar systems comparable to our own Milky Way system, with about the same range of dimensions and containing roughly the same number of stars as the local group. Compare with the caption to Plate 3. (*Photo:* Tautenburg)

7   This version of the Hertzsprung–Russell Diagram was published by H. N. Russell in 1914. Hertzsprung had already plotted a similar diagram, but with fewer stars, in 1907. It will be noticed that the distribution of the points on these diagrams is relatively wide; there are two reasons for this: in the first place only comparatively few stellar distances were then known, and secondly the absolute magnitudes as calculated from the distances were not yet sufficiently exact. Only the main trends can be recognised, the grouping of the giants in a horizontal branch, and the dwarfs in a diagonal band stretching from the top left-hand side of the diagram to the bottom right. The existence of white dwarfs is indicated.

8   This diagram combines the colour-magnitude diagrams of open and globular star clusters. Within each cluster there is only a narrow range of possible stellar conditions. In their lower sections the main sequences of all clusters coincide; in their upper portions, however, they branch off more or less abruptly to the right, and terminate at luminosity maxima which are characteristic for each cluster. This trend in the colour–magnitude diagrams can be explained if one assumes that the clusters are groups of stars of roughly the same ages; stellar evolution follows the path described in figure 13. The stars at the upper end of the main sequence use up their hydrogen fuel much more rapidly, and so leave the main sequence sooner. Hence the

place where the main sequence breaks off is indicative of age. In the youngest cluster ('*h*' and '*χ*' Persei aged 4 million years) stars have only just begun to leave the main sequence; in the eldest (Messier 67 aged between 4,000 million and 5,000 million years) all stars of greater absolute magnitude than the Sun have already left the main sequence.

The colour–magnitude diagrams of the globular clusters, of which Messier 3 and Messier 92 are examples, occupy mainly the region of the yellow and red giants; they are estimated to be at least 6,000 million years old. [(*Source: Meyer's Handbuch über das Weltall* (according to Sandage)]

9 Typical stellar spectra which were photographed by G. Miczaika and K. Bahner with the 72-cm. reflector of the observatory on the Königsstuhl near Heidelberg. In the spectra of A-type stars the hydrogen lines of the Balmer Series predominate; in subsequent types the Fraunhofer lines H and K of calcium become more pronounced. In general the wealth in lines becomes greater with advancing spectral type (diminishing temperature); in stars of the lowest temperatures apart from the lines which are due to atoms molecular bands also appear.

10 This schematic colour–magnitude diagram (1955) was taken from *Meyer's Handbuch über das Weltall* which in turn was based on an outline in *Grundriss der Astrophysik* by H. Siedentopf. It marks the areas in which the various types of stars lie on the diagram, stars about whose physical characteristics, in particular their luminosities and effective temperatures, reasonably reliable assertions can now be made. Luminosities are given in terms of the Sun (cosmic standard candle); the Harvard spectral classes B to M correspond to the colours blue–white–yellow–red in continuous sequence. The relation between spectral type, colour index and effective temperature is shown in the following table:

133

SPECTRUM	COLOUR INDEX	EFFECTIVE TEMPERATURE
B 0	-0.2	25,000°K
A 0	0.0	10,700
F 0	$+0.4$	7,500
G 0	$+0.6 (+0.8)$	6,000 (5,200)
K 0	$+0.9 (+1.2)$	5,100 (4,200)
M 0	$+1.5 (+1.8)$	3,400 (3,400)

Figures in brackets refer to giants.

[*Source: Meyer's Handbuch über das Weltall* (according to Sandage)]

11 The Pleiades are a galactic cluster consisting of not much more than 100 stars, and of relatively recent origin (about 80 million years). The cluster is embedded in a vast cloud of nebular matter. The extremely short-wave radiations of bright B-type stars cause the filament-like nebular material to glow (emission nebula). (*Photo:* Tautenburg)

12 The great Orion Nebula is an example of the interaction between stars and interstellar matter. The chaotic nebular masses which are in a state of extreme turbulence are either made to glow by the hot B-type stars embedded in them, or they reflect the radiations of A-type and F-type stars; alternatively they stand out as dark clouds silhouetted against the general background of Milky Way stars. Within the cloud-like agglomerations of nebular masses stars are forming in small groups (associations) through disintegration of larger condensations (globules?). The so-called Trapezium in the interior of the nebula—not visible on this plate owing to the long exposure—forms the core of one such association whose stars are hardly likely to be more than 2·5 million years old. (*Photo:* Tautenburg)

13 The life-cycle of a star of the mass of the Sun shown on the H–R diagram (according to Martin Schwarzschild). The

star which formed out of condensation of interstellar matter passes through the contraction stage—from right to left—in a relatively short time (a few million years or so), until such time as the hydrogen fusion reactor in its interior releases energy. So long as these nuclear processes continue luminosity and effective temperature undergo little change, and the star maintains its place on the main sequence. When the store of hydrogen fuel has been largely depleted, and a core consisting principally of helium has formed, the helium reactor brings about a raising of the central temperature and an increase in diameter; the path on the H–R diagram leads upwards to the right. Further evolution, after the helium has been used up, is still uncertain, but most probably, after a period of unstable conditions (variable star), leads to a collapse (nova?), finally to terminate as a white dwarf. (*Source: Stellar Evolution*, Otto Struve)

14 Cosmic evolution: Otto Struve described the following pattern of interaction between diffuse matter and stars in the course of the evolution of the Galaxy. In this pattern he tries to relate the individual cosmogonistic phenomena— formation of stars, double stars, planetary systems and stellar systems—in a sort of sequence. The picture is undoubtedly extremely hypothetical, but it does seem to explain a number of observations. (*Source: Stellar Evolution*, Otto Struve)

Further Reading

ABETTI, G. *The Sun*. (Translated J. B. Sidgwick), Faber and Faber, 1957.

ABETTI, G. *Stars and Planets*. (Translated Dr. V. Barocas), Faber and Faber, 1966.

ABETTI, G. and HACK, M. *Nebulae and Galaxies*. (Translated Dr. V. Barocas), Faber and Faber, 1964.

BERGSØE, P. *The Universe and Man*. (Translated K. John), Methuen, 1962.

BONDI, H. and others. *Rival Theories of Cosmology*, Oxford University Press, 1960. (Based on talks on the Third Programme, 1959.)

GAMOW, G. *A Star Called the Sun*, Macmillan and Co. (London), 1965.

PLEDGE, H. T. *Science Since 1500* (Chapters XXI and XXII), H.M. Stationery Office, 1966.

STRUVE, O. and ZEBERGS, V. *Astronomy of the 20th Century*, Macmillan and Co. (London and New York), 1962.

Index